AF606423

The Dwarf Novae

THE DWARF NOVAE

by John S. Glasby, B.Sc., F.R.A.S.

Constable London

First published in 1970
by Constable & Co Ltd
10 Orange Street London WC2

Printed in Great Britain
by The Anchor Press Ltd, Tiptree, Essex
ISBN 0 09 457140 6

To Janet

Contents

Preface

The dwarf novae, although only a small class of variable star, are amongst the most interesting to have been discovered and the study of their light and spectral variations is a relatively recent one when compared with the many centuries during which the science of astronomy as a whole has attracted the attention of thinking men since the dawn of civilisation. Indeed, only three members of this class were known before the present century and it is only during the last decade or so that precise measurements using the largest instruments in operation have provided us with an insight into the physical nature of these variables.

Since many problems still await solution, the discussion is necessarily incomplete but it includes most of the important questions concerning which reasonably definite statements can be made at the present time. The subject is presented mainly from the observer's point of view for two reasons. Firstly, the writer is primarily an observer and secondly it is only in recent years that theoretical considerations of the dwarf novae have achieved prominence. Within these limitations, the book is believed to furnish an up-to-date picture of our present knowledge of these stars.

It is unfortunate that the large majority of these variables are extremely faint objects, even at maximum brightness, and their outbursts are totally unpredictable making it necessary to observe them on every possible occasion if complete records of their light variations are to be obtained. The inevitable result of this is that reasonably detailed light curves are available for only a small number of stars and by far the greater number are not being observed at all. This will be immediately obvious from the small number of light curves which are given in Part Two of the book in which the characteristics of the individual dwarf novae are listed. Here it should be mentioned that several of these light curves are drawn from observations made by the author over the last four years and are necessarily far less accurate than would have been the case had more observations been available. The author makes no apologies for including these since not only do they give some indication of the behaviour of these particular variables but it is also hoped that they will stimulate other observers with moderately large instruments to take up this observational work and provide much needed data. The writer is responsible for those light curves which are not specifically attributed to others and in this connection, he gratefully acknowledges his indebtedness to the University of Chicago Press for permission to include several of the diagrams from the *Astrophysical Journal*, the British Astronomical Association, the American Association of Variable Star Observers and the Variable Star Section of the Royal Astronomical Society of New Zealand for the use of many thousands of observations made by their members over the past seventy years.

Stevenston, Scotland
February 1969

J.S.G.

Introduction

The study of variable stars may be said to have begun in 1572 with Tycho Brahe's observations of the brilliant galactic supernova in the constellation of Cassiopeia, although it gathered little impetus until a century or so later. In 1596, Fabricius noted the peculiar light variations of Mira, variations shown to be periodic in nature forty-two years later by Holwarda. Much later, in 1782, Goodricke provided the correct explanation of the sudden decreases in the brightness of β Persei (Algol) which had been detected by Montanari in 1669, suggesting that here we have a close binary system consisting of two stars of very different luminosities which mutually eclipse each other in the course of one revolution around their common centre of gravity.

In spite of these early observations, it was not until the middle of the nineteenth century that the study of variable stars was put on a systematic basis by the work of Argelander – and with the subsequent development of photographic methods, this particular field of astronomical investigation has expanded so rapidly that at the present time almost 25,000 variables are now known, their continued observation having made it possible to divide them into several well-defined classes based mainly upon their light curves and to a lesser extent, their spectra.

Although the dwarf novae, with which this book will be concerned, form only a very small class of variable star and were completely unknown a little more than a century ago, they are now gaining increasing importance in the field of astronomical research, not only by virtue of their light and spectral features but because they present the astronomer with a wide variety of problems, many of which are, as yet, unsolved. The main aim of this book is to present the reader with an overall picture of our present knowledge of these peculiar stars, indicating the various problems still under active investigation. It would be impossible, however, and unwise, to regard the dwarf novae in isolation. In several aspects, they show remarkable similarities to other types of variables, particularly the recurrent novae and the W Ursae Majoris eclipsing systems.

As with all other branches of astronomy, the investigation of variable stars follows two main paths; the first, quite naturally, is the observational approach while the second is purely theoretical. The observer accumulates a mass of data (very often a long and tedious process) and attempts to find, usually by graphical means, correlations between the various parameters. Sometimes, the relationships which are discovered are extrapolated into hitherto unknown regions and theories are developed which stand or fall according to whether future observations agree or disagree with these hypotheses.

In the case of variable stars, for example, we may plot the apparent visual or photographic magnitude against time thereby obtaining the light curve which, in turn, may be compared with the spectroscopic changes occurring in the star through the course of each light cycle. At this point, the theoretician usually enters the field and puts forward various models in an attempt to explain the observations. Only for one particular class of variable is it possible for us to be certain that the model which has been advanced is the correct one. Such variables, typified by Algol and β Lyrae, are known to be eclipsing systems and both the light and spectroscopic variations can be explained quite satisfactorily on the basis of two stars revolving about a common centre of gravity, the plane of the orbit lying close to, or in, the line of sight as seen from Earth, so that the components mutually eclipse each other during one revolution.

In most other cases, the theories which have been put forward are plausible and explain several of the features observed. The Cepheid and long period variables appear to be pulsating stars and a great deal is already known about the mechanism of such pulsatory changes in these variables. The R Coronae Borealis variables are supergiant stars containing an over-abundance of carbon and its compounds in their outer atmospheric levels which at times condenses to form an obscuring layer thereby reducing the amount of light we receive; while the novae and supernovae are exploding stars which throw off an outer layer of hydrogen or are almost completely disrupted at the time of an outburst.

Actually, the models suggested above are all very much over-simplified. There is still a great deal we do not know about these variable stars, information which will doubtless be obtained from future investigations. Because of the incomplete state of our knowledge at the present time there are, inevitably, several opposing theories in vogue for explaining the many peculiarities of each type of variable. Modifications, some minor, others of a more major character, will have to be made as more observational data accumulates but gradually the boundary between the known and the unknown is being pushed back further and further as techniques become more refined and larger instruments are brought into operation.

As with all other sciences, astronomy uses its own particular vocabulary, some of the terms of which will be unfamiliar to the layman and it will be worthwhile to explain certain of these insofar as they relate to variable star observation.

Designation of Variable Stars

It was soon recognised that variable stars form a distinct and important class and consequently there had to be some ready means of differentiating them from the non-variable variety. A few, of course, are naked-eye objects and were given Greek letters by Bayer during his charting of the constellations and these have since been retained. For example, the star discovered in Cetus by Fabricius and now known to be the prototype of the long period variables, was classified as o Ceti by Bayer who observed it when at maximum brightness and was unaware of its variability. By far the vast majority, however, are telescopic objects and it was universally agreed to use capital Roman letters for them beginning with R, followed by S, T and so on to Z in each constellation.

The series then continues with RR, RS to RZ; SS to SZ and so on until ZZ is reached when we go on with AA, AB to AZ, BB to BZ, etc., until the sequence ends with QZ. This provides us with a total of 334 combinations and in those constellations where the number of variables exceeds this number, subsequent stars are designated V335, V336, and so on. It will be noted that the single letters from A to Q have not been used. The reason for this is that these letters had already been employed for non-variable stars in the newly-discovered southern constellations.

Stellar Magnitudes

Throughout this book the reader will encounter several different kinds of magnitudes, all of which have their own special meaning. Essentially, the term *magnitude* gives a measure of the luminosity of a star; the fainter the star, the larger is the numerical value of the magnitude. It is only on the minus side of magnitude 0.0 that this is reversed, the figure here increasing with the brightness. Apart from some novae and supernovae at maximum, only two stars have negative magnitudes; Sirius with an apparent magnitude of −1.58 and Canopus which is of apparent magnitude −0.86.

(a) Apparent magnitude: This is the brightness determined directly by the eye or the photographic plate and takes no account at all of the distance of the star and therefore of its actual luminosity. We may measure it in a variety of ways, all of which have the common property that the magnitudes are proportional to the logarithms of the luminosities and not to the luminosities themselves. The classification employed by Ptolemy in his *Almagest* was a division of the stars visible to the naked eye into six groups of roughly equal ratios of brightness. This scheme made the first magnitude stars approximately one hundred times brighter than the sixth and formed the basis for the modern system due to Pogson who, in 1856, introduced the concept of making the first magnitude stars exactly one hundred times brighter than the sixth. Each magnitude step is therefore a luminosity ratio of 2.512. Naturally there must be a zero point for such a sequence and by international agreement this has been defined as the mean magnitude of certain A0 type stars between 5.50 and 6.50 magnitude in the North Polar Sequence (a list of accurately-measured stars lying close to the north celestial pole).

When the eye alone is used, we obtain visual magnitude which, together with the closely-related photovisual magnitudes (obtained photographically by means of a colour filter and special isochromatic plates adjusted to a light sensitivity similar to that of the eye), are yellow magnitudes. Ordinary photographic magnitudes on the other hand, are blue-violet ones since the photographic emulsion usually employed is more sensitive to the wavelengths in the blue-violet region of the visible spectrum than to those in the red. As a result, photographic magnitudes are brighter than visual ones for blue stars and fainter for red ones. On the Harvard scale, stars of spectral class AO have identical visual and photographic magnitudes. The difference between the visual and the photographic magnitudes is known as the colour index of a star. In general, colour indices are quite small, usually lying between −0.5 and +2.0 magnitudes. Only rarely, as in the case of certain N type stars (which are very red and nearly all variable), do we encounter a colour index in excess of +3.0 magnitudes.

Three other classes of apparent magnitude are often encountered in variable star investigations. The photometric magnitude is the visual magnitude as estimated by means of a photometer and is, of course, far more accurate than when determined by the eye alone.

The photoelectric magnitude is met with when very small and rapid fluctuations have to be measured. Here a photoelectric cell is used, an instrument which will be described in detail later in the book. Since individual cells have different sensitivities to the various wavelengths there is no photoelectric scale, such results being given as the difference in magnitude between the variable and a nearby star of constant and comparable magnitude.

Finally, we have the bolometric magnitude which is a measure of the total radiation emitted by the star taking into account not only the visible light but also the infra-red and ultra-violet radiation, including that which is absorbed by our own atmosphere. In this case, visual and bolometric magnitudes agree for stars of type Go, very similar to our sun.

(b) Absolute magnitude: The apparent magnitude of a star which we have just discussed, no matter by what means it is obtained, gives us no indication whatsoever of the intrinsic luminosity of the star. A faint star which is close to us and a highly luminous one at a great distance may both be of the same apparent magnitude. The two first magnitude stars in Orion are good examples of this; Rigel, slightly the brighter of the two being about four times more distant than Betelgeuse.

We must therefore have some means of measuring the true brightness since this is obviously an important function of a variable star. Such intrinsic luminosities are indicated by the absolute magnitudes (M) as opposed to the apparent magnitudes (m). The definition of the absolute magnitude of a star is really quite a simple one. It is merely the apparent brightness of the star if it were situated at a standard distance from us, this distance being 10 parsecs (32.6 light years). At any other distance the value (m−M), which is known as the modulus of the distance, and which is clearly a function of the distance, is expressed by the simple relationship.

$$\log D = 0.2\,(m-M)+1.0 \qquad (1)$$

where D is the distance in parsecs. If we measure the distance in light years, the constant on the right hand side of the above equation becomes 1.513. Quite obviously, we can always measure m and we only require to know either D or M in order to be able to calculate the other. As we shall see, this simple relation is of great importance in deriving the distances of the dwarf novae. Powerful statistical methods of determining absolute magnitudes for the different types of stars, both giants and dwarfs, have been developed to a degree which permits quite accurate estimates to be made.

Spectral Types

If we were forced to rely entirely upon measurements of the light variations of variable stars our knowledge of their physical characteristics would be severely limited indeed and any theory advanced to explain such changes in brightness would be, at best, mere speculation. Fortunately we have another very powerful tool at our disposal, namely the spectroscope. It is perhaps well to remember that, unlike the chemist or physicist who is able to make detailed examinations of objects at first hand, the astronomer must work only with the light, often extremely feeble, which comes to us from the stars across distances of many billions of miles. There is no way at all by which he can change the light and mould it to his own fashion to provide more information than is originally contained therein.

Fortunately, not only does the spectrum provide us with a good idea of the surface temperature and the physical and chemical constitution of a star but it has given us conclusive proof that many, if not all, of the dwarf novae are binary systems, consisting of two stars of very different spectral classes.

Before discussing the classification of stellar spectra, we must first of all mention the three basic types of spectra which may be obtained either by passing light through a prism or by use of a diffraction grating. The spectrum we get from an incandescent solid or liquid (under certain circumstances also from a glowing gas) is *continuous*, being merely the familiar band of rainbow colours, each blending imperceptibly into the next and showing neither bright nor dark lines across the spectrum.

Under the more usual conditions encountered with a glowing gas, the spectrum consists of a series of bright lines superimposed upon a dark continuum, this being known as an *emission* spectrum.

The most common type of stellar spectrum, however, is an *absorption* spectrum in which the continuum is crossed by dark lines which are due to the light from the much hotter lower levels passing through the cooler gases of the star's atmosphere. Now every chemical element (and certain compounds) in the stellar atmosphere absorbs those particular wavelengths which are peculiar to it, thereby leaving its fingerprint in the spectrum. We are therefore able to tell not only which elements are present but also, from the intensity of the various lines, their relative abundance and something about the physical conditions which prevail in the atmosphere.

An important question now arises. We know from laboratory experiments that both temperature and pressure will affect the spectral lines, but which of these two parameters exerts the predominant effect in stellar spectra? The answer may be determined by two methods. Either we examine the spectra of gases in the laboratory under varying conditions of temperature and pressure or we study the spectra of many different types of stars using independent methods of measuring the prevailing temperatures and pressures. Both ways yield essentially the same result. By far the greater effect is produced by temperature and not pressure. This is an important conclusion as far as the dwarf novae are concerned for, as we shall see, one of the components of these binary systems is a fairly typical white dwarf star, a peculiar class of object possessing an extremely high density.

Classification of Stellar Spectra

Not long after the application of the spectroscope to the study of the stars, Secchi divided stellar spectra into four fairly well defined groups, a fifth being added by Vogel. Secchi's classification is no longer used, having been superseded by the Harvard sequence of stellar spectra first introduced by Pickering around 1890. The spectral classes were originally arranged in alphabetical order from A to Q but as more information was gathered, certain letters were omitted and other modifications became necessary until now the sequence bears little resemblance to the original.

The present sequence: W, O, B, A, F, G, K, M, N, R, S and Q (used for novae) is in order of decreasing surface temperature although there is very little difference among types M, N, R and S, all having temperatures in the range 2,000°K to 3,000°K. Almost all of the stars in these four spectral classes are variable.

In the case of variable stars, the spectrum given is that of the star when at maximum brightness, unless otherwise stated. Naturally there are certain changes in the spectrum as the star fades (or brightens) and it has proved necessary to standardise things in order that comparisons may be made. It is fairly obvious, too, that further subdivisions must be made in the above classes and to this end the numbers 0 to 9 are used as suffixes. For example, a star of type F2 will be two-tenths of the way from type F to the following class G.

Various additional symbols are used to denote peculiarities but in this book we shall confine ourselves to the following which are used extensively to describe the spectra of the dwarf novae –

The prefix 'd' indicates that the star is a dwarf, the enhanced lines in the spectrum being generally weak and the calcium lines usually strong. Similarly, 's' indicates that the spectral lines are sharp but that the star does not belong to the giant class (sharpness of the lines is usually associated with giant and supergiant stars). Lastly, we shall use the prefix 'e' to denote that bright emission lines are present in the spectrum, either throughout the entire light cycle or during certain phases of the light variation.

Luminosity Classes

Observations which have been recently carried out at Yerkes Observatory, especially by Morgan, have led to the recognition of five classes of luminosity among stars. Although the spectral class may be determined with little difficulty from an inspection of the spectrum, it is necessary to establish certain criteria before the luminosity class may be determined. Basically, there are two such criteria, namely –

(a) Owing to the much lower densities of the giant and supergiant stars compared with those of the dwarfs, the mean free path of an electron is far longer in the former and consequently we would expect ionised atoms (those which are deficient of one or more electrons in their outer shells) to be more abundant in the former than in the latter compared with neutral atoms. This being the case, the lines due to ionised atoms will be enhanced in the spectra of the giant stars.

(b) The lines due to any particular chemical element will be much narrower in the spectrum of a supergiant star than in that of a dwarf and it may be shown in this case that the fineness of the lines is related directly to the extremely low pressures in the outermost layers of these huge stars.

From an examination of the stellar spectrum therefore, particularly from high-dispersion spectrograms, both the spectroscopic type which provides us with a knowledge of the surface temperature, and the luminosity class which indicates whether the star is a dwarf, giant or supergiant, may be determined with quite considerable accuracy. The importance of the luminosity class of a star is at once apparent when we consider that an M-type star, for example, may be either a red dwarf, much smaller and therefore considerably less luminous than the sun – or a red supergiant many thousands of times more luminous.

The Equipotential Surface and Inner Lagrangian Point

As we shall see in the ensuing chapters, it is now generally believed that the dwarf novae are close binary systems in which a stream of gas, mainly hydrogen, flows from one component to the other. In order to gain an understanding of the dynamics of these systems, it will be necessary to introduce certain terms which are unfamiliar to the layman. Of these, the equipotential surface and inner Lagrangian point are among the most important and here we shall discuss these in some detail.

The simplest case of all is that of a single star which is in the process of expansion, as is the case when such stars begin evolving to the right of the main sequence. Now the gravitational field of such a star will limit the expansion since all particles in the star are acted upon by such a gravitational field. In this particular instance, as the field extends equally in all directions, the equipotential surface will take the form of a sphere, this being the distance at which all particles will have zero velocity. In other words, the mass of the star will be contained within the equipotential surface (although it may not completely fill it since this represents the point at which the fastest-moving particles attain zero velocity). Naturally the diameter of the equipotential surface will vary greatly from star to star, being dependent mainly upon the mass and the temperature.

When we come to consider a close binary system, the picture is a little more complex. Instead of two spheres surrounding each component as we might expect, the configuration is dumb-bell shaped (Fig. 1), the two surfaces being joined at a point which is known as the inner Lagrangian point.

The importance of the inner Lagrangian point in the study of the dynamics of close binary systems such as the dwarf novae lies in the fact that when one of the components begins to evolve away from the main sequence (nearly always to the right), this star will commence to expand until it completely fills its lobe of the equipotential surface. Once this happens, further evolution will demand that the star should continue to expand with the result that mass must escape from within the confines of the equipotential surface. Now where is this mass most likely to escape? Clearly through what we may term the point of least resistance. Quite obviously, from an inspection of Fig. 1, this will be through the inner Lagrangian point for it is here that the gravitational attraction of the companion star is greatest. Since it is unlikely that the companion star (in the case of the dwarf novae we shall see that this is a fairly typical white dwarf) completely fills its lobe of the equipotential surface, this stream of gas will then take up an orbit around the smaller component.

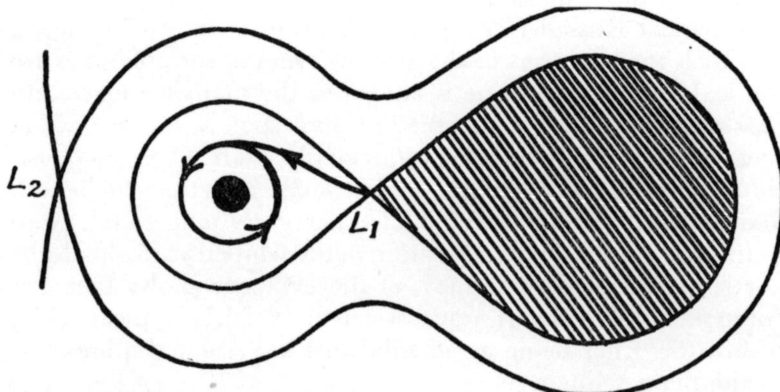

Fig. 1 – The equipotential surface of a close binary system. L_1 and L_2 represent the inner and first external Lagrangian points respectively.

In succeeding chapters, we shall see that such a model has been suggested by several astronomers for AE Aquarii and many other dwarf novae and explains several of the characteristics of these stellar systems.

Mention must also be made of the first external Lagrangian point since it is here that any mass loss from the system into space takes place. When we come to discuss the precursors of the W Ursae Majoris variables which are, themselves, thought by many astronomers to be the congenitors of the dwarf novae, we shall find that there is a lot of evidence available which suggests that an appreciable proportion of mass is lost from these evolving systems through the first external Lagrangian point thereby influencing the evolution of such stars.

The Russell-Hertzsprung Diagram

There are several parameters of the stars as a whole which can be fairly accurately determined, either directly or indirectly and the first attempt to introduce some order into the rather bewildering array of data accumulated was made some fifty years ago by Russell and Hertzsprung working initially independently of each other. Although the type of diagram originally constructed by them has been considerably modified in recent years, it still remains the basis on which the current theories of the evolution of the stars have been built.

A typical Russell-Hertzsprung diagram is shown in Fig. 2, in which the positions of the various types of variable stars are marked. The straight line is known as the main sequence and represents the position occupied by stars such as the sun which have not yet begun evolving appreciably; stars whose composition is virtually all hydrogen with a small percentage of helium and an even smaller amount of other gases and metals (also in the gaseous form). The position of the white dwarfs should be particularly noted since most of the evidence shows that one component of a dwarf nova system is a fairly typical white dwarf, the other being a red subdwarf star situated quite close to the main sequence.

As far as our own galaxy is concerned, it is not easy to determine the absolute magnitudes for many stars since this requires a knowledge of their distance (not always an easy parameter to determine). The situation is much simpler when we come to construct the Russell-Hertzsprung diagrams for the stars in the globular clusters or the Magellanic Clouds since here we may make the assumption that all of the stars are essentially at the same distance from us and their absolute magnitudes may be derived directly from their apparent magnitudes. When this is done, we find such diagrams to be almost identical to those obtained for the galactic stars.

As will be readily understood, owing to the inherent inaccuracies in determining the absolute magnitudes of stars within our own galaxy, the original diagrams suffered from the disadvantage that the zones into which the various types of stars fell were very wide and diffuse and it is only recently, as more measurements with an improved accuracy are being obtained, that much narrower zones, although still with a scatter of one or two absolute magnitudes on either side, are being obtained.

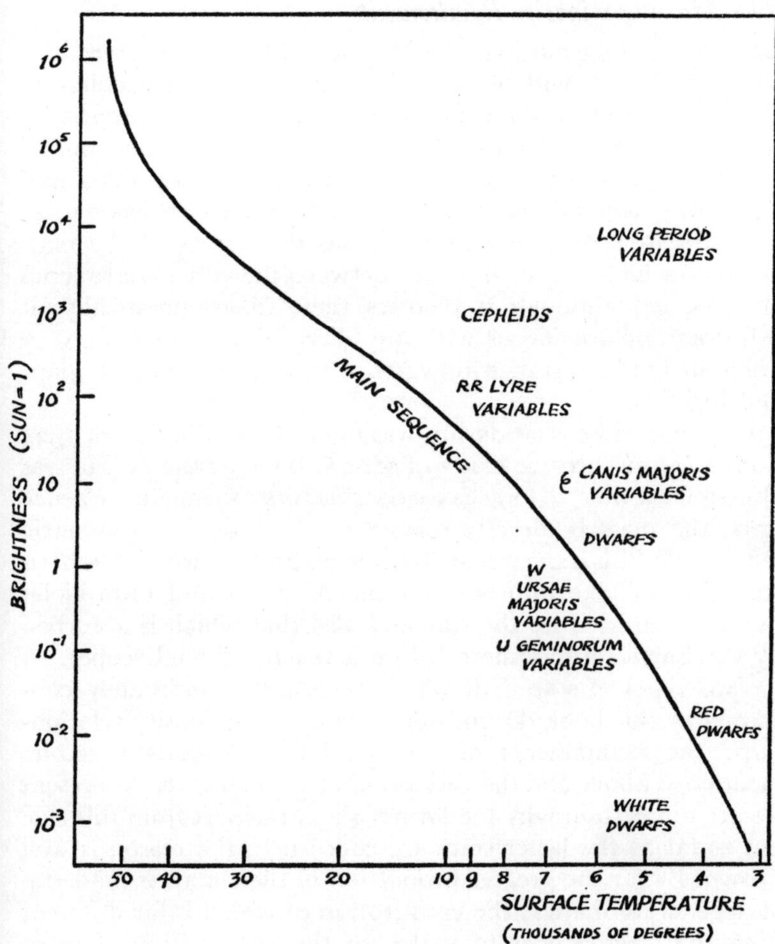

Fig. 2 – A typical Russell-Hertzspring diagram illustrating the positions of the various types of variable stars.

In addition, there is now a tendency among astronomers to replace the spectral classification of the abscissa of the Russell-Hertzsprung diagram with one based upon the colour index.

The reason for this is that such colour indices can be measured without any ambiguity and with a much greater degree of accuracy.

The Mass-Luminosity Relationship

During the discussion on the W Ursae Majoris variables in a later chapter, it will be seen that these stars are peculiar in that they do not obey what is known as the mass-luminosity relationship. In order to fully understand the implications of this, a brief mention of the relationship between mass and luminosity will not be out of place here. As we have seen, the Russell-Hertzsprung diagram has demonstrated the existence of a high degree of order between the very diverse types of stars, yet although it pictures these diagrammatically, it still does not provide us with any direct link among the very different kinds of star, subdwarfs, white dwarfs, dwarfs, giant and supergiants.

One very close relationship was found by Eddington in 1924 from a comprehensive study of several binary systems. He was able to show that, so long as we consider only the main sequence stars, the mass is directly related to the absolute bolometric magnitude (this is the visual absolute magnitude which has been corrected to take into account the infra-red and ultra-violet radiation emitted by the star and also that which is absorbed by the Earth's atmosphere before it reaches the telescope).

Two types of star with which we shall be intimately concerned in this book do not obey the mass-luminosity relationship; the components of the W Ursae Majoris variables mentioned above and the very peculiar white dwarfs. At present we are not certain why the former should depart from this rule but as far as the latter stars are concerned, the reason is well known. By far the greater proportion of their mass is made up of degenerate matter, the constitution of which is far different from that which goes to make up the composition of more normal stars. This peculiarity of the white dwarfs will be discussed in detail in a later chapter.

PART ONE

1. The Discovery of the Dwarf Novae

Whenever we look at the stars spread out over the vastness of the heavens, one fact is immediately obvious, namely that from the brightest down to the faintest which can be seen, every intermediate grade of brilliance is represented. This gradation in luminosity continues all the way down through the telescopic range until we reach a point where long exposure photographs using the largest telescopes available record the faintest objects of all known to the astronomers. The classification of the stars into magnitudes originated more than two thousand years ago with Hipparchus who divided those stars visible to the naked eye into six groups, or magnitudes – a classification which has survived to this day except for a minor modification made, as we have seen earlier, by Pogson.

Hipparchus and all of the early astronomers regarded the brightness of the stars as an unalterable feature and for almost two thousand years this view remained unchallenged. The heavens were considered to be unchanging and unchangeable. Then, in 1572, a spectacular star of exceptional brilliancy appeared in the constellation of Cassiopeia, so bright that it was visible even in the daytime, rivalling Venus, the brightest of the planets. After a few months, this new star commenced to fade until early in 1574 it was no longer visible to the naked eye. Some years later, in 1596, the peculiar changes in brightness of another star in Cetus were noticed by Fabricius. Although this particular star did not attain anything like the brilliance of that of 1572, its discovery may be regarded as being more important since, in 1638, Holwarda announced that the light variations of this star are periodic. Thus was born the branch of astronomy which deals with the study of variable stars, those whose brightness is not constant but varies with time.

At first, only a few such stars were discovered but later, with the introduction of celestial photography, progress in this field accelerated dramatically and as more data were accumulated it became obvious that these stars could be divided into several different and distinct classes depending mainly upon the nature of their light fluctuation. By the middle of the nineteenth century, the main classes which had been distinguished were the long period variables, semi-regular and irregular variables, the Algol and β Lyrae eclipsing systems, the Cepheids and the novae. The form taken by the light curves of these types of variable stars was reasonably well known. As this was some time before the application of the spectroscope to the study of variables, all of the knowledge of these stars was obtained simply from a study of their light curves. It was not until several decades later that the spectroscope opened up fresh avenues of research and enabled astronomers to provide various models for the different types of variable stars.

The discovery of the first dwarf nova came more than a century ago on December 15, 1855 when John Russell Hind[1] observed a new star in the constellation of Gemini. At the time of its discovery it was a blue-white object of the ninth magnitude in a part of the sky which was well known to Hind who had spent many years searching the zodiacal constellations for asteroids. Three days later, the star was at least half a magnitude fainter and by the end of the year had faded below the limit of Hind's instrument.

The fact that this was a new type of variable star was not at first realised. The behaviour of this star which Hind described as having a planetary appearance, strongly suggested that it might be a nova – a star which suddenly increases in brightness by between six and twelve magnitudes and then fades somewhat less rapidly after remaining at maximum brilliance for a matter of a few weeks or so. In the following March, however, it was again seen at maximum, this time by Pogson[2] and this second appearance immediately disproved the nova theory since these stars are seen at maximum only once (present theories of the nova phenomenon suggest that a star may undergo a nova explosion several times during its lifetime but the interval between outbursts is of the order of several centuries). The recurrent novae which have been recognised only during the present century also have mean periods of many years.

The unusual behaviour of this star resulted in it being kept under constant observation by several observers and subsequent maxima recorded during the succeeding years soon established beyond any doubt that it was completely unlike any of the variable stars known at that time.

The long period variables, several of which were known and had been closely studied, have mean periods between 100 and 700 days which is of the same order as that of U Geminorum as this new variable was designated. All of the former, however, are red stars even at maximum and their light changes are relatively slow, such variables taking several months to rise from minimum to maximum.

Sufficient was known too of the light variations of the Cepheids and eclipsing variables which are characterised by an almost clockwork regularity in their period which is completely absent in Hind's variable, a star distinguished by abrupt and quite unpredictable increases in brightness amounting to about five magnitudes and occurring at intervals of a few months with more or less smooth minima between outbursts.

For more than forty years U Geminorum remained virtually unique among variable stars. Then, in 1896, a second of these peculiar variables was discovered by Miss Wells[3]. As it turns out this particular star, SS Cygni, is the brightest of all the dwarf novae, reaching 8.1 magnitude during its most brilliant maxima and rarely falling below 12.1 magnitude at minimum. As we shall see in the next chapter when we come to discuss the light variations of these stars in detail, SS Cygni differs from U Geminorum in that three distinct types of maxima have been recorded; long, short and anomalous, only the first two kinds being known in the case of the prototype star.

It is convenient at this stage to say a few words about one star which was observed by Peters[4] as long ago as 1865, only ten years after the discovery of U Geminorum. This is the peculiar variable T Leonis. When seen by Peters, it was as bright as 10.1 magnitude but it soon faded into invisibility in a manner typical of the dwarf novae and has been seen on only very rare occasions since then at anything like its maximum brightness. Quite naturally, if we had only its light curve to go on we would be perfectly justified in classifying this star among the novae although admittedly its amplitude of 5·3 magnitudes is somewhat smaller than that of a typical nova. When at minimum it is an inconspicuous object of 15·4 magnitude requiring large apertures for its observation. Even as recently as 1948, it was thought to be perhaps constant at its minimum brightness[5]. Since that time, however, its spectrum at minimum has been examined in detail, particularly by Kraft[6] who has shown that at this phase the spectrum is very like those of many other dwarf novae, consisting of bright emission lines of hydrogen, helium and ionised calcium. It is now generally accepted by most astronomers that T Leonis belongs to the U Geminorum class of variables although it is far from being a typical member and as yet we know nothing of its mean period except that it appears to be very long indeed.

During the first decade or so of the present century, further representatives of this class were found, most of them being initially mistaken for novae. RU Pegasi and Z Camelopardalis were both discovered in 1904 and RX Andromedae in the following year. Two years later, SS Aurigae and X Leonis were added to the class followed by SU Ursae Majoris in 1908, SW Ursae Majoris in 1909, UV Persei in 1911 and TZ Persei in 1912.

Quite clearly, the dwarf novae do not constitute a class as numerous as, for example, the long period variables or the Cepheids. Perhaps, as far as actual numbers are concerned, it would be fair to compare them with the galactic novae which have been discovered to date. Once celestial photography was introduced to survey the heavens, several more were found on patrol plates taken at short intervals, being generally recognised by their characteristic light variations. As we shall see in Part Two when we come to discuss the individual members of this class, there is one difference between the dwarf novae and the ordinary. Unlike the novae, the large majority of these variables are extremely faint objects, even when at maximum, making their discovery and positive identification much more difficult. There is also the added complication that their stay at maximum is normally quite short, often being a matter of only one or two days, and as a result it is very probable that several more still await discovery. When a faint variable is discovered on plates taken of regions well away from the galactic plane and has a light curve throughout the maximum typical of that of the dwarf novae, it is reasonable to assume that it is a member of this class although positive identification will only come when further maxima are observed. The problem is a little different, however, when those areas close to the Milky Way are surveyed since a large number of faint novae have been found here (these stars being concentrated along the boundaries of the Milky Way) and as a result it is somewhat more difficult to assign such a star to the dwarf novae.

So far, we have treated the dwarf novae as being essentially one distinct class of variable star with perhaps some minor variations in the light changes but nothing of a major character which would suffice to make us subdivide the group into more than one basic type. Indeed, several years of continuous study of these stars were to pass before it was recognised that not all of these variables could be fitted into the same group as that typified by U Geminorum and SS Cygni. Most of the dwarf novae, it is true, are found to fade fairly smoothly to a normal minimum after each outburst but a small number have been found which, after certain maxima, remain at some intermediate brightness for an unpredictable period which may last anything from a few days to many months. Such periods of comparative inactivity are known as 'standstills' and the stars of this small subgroup are known as the Z Camelopardalis variables after the first member of the class shown to exhibit this phenomenon.

The two groups are not, of course, easy to distinguish from each other. There are two main reasons for this. It will be appreciated that in many instances, the fainter dwarf novae have not been observed continuously for a sufficiently long period for the presence of such 'standstills' in the light curve to be conclusively demonstrated. Secondly, as we shall see in later chapters, the frequency with which these 'standstills' occur varies quite considerably from one star to another. In some, they are quite a common feature while in others only one or two have been recorded during the entire period over which the star has been under observation. In a situation such as this, it is inevitable that future data will result in the transfer of some of those stars at present classed as U Geminorum variables into this much smaller subgroup of the Z Camelopardalis stars.

Since the introduction of photographic techniques and particularly the many surveys which have been made in recent years of selected Milky Way fields in the search for other types of variable stars, the number of known dwarf novae has steadily increased and at the present time about 150 stars have been classed as known or suspected members of this kind, the great majority being of the U Geminorum class, only a small handful belonging to the Z Camelopardalis subgroup.

For the amateur astronomer, on whom the task of keeping these peculiar variables under constant observation mainly falls, there is still much to be done in the way of the discovery of new variables of the dwarf novae type. Being so faint at all phases of their light variation, they are not, of course, the easiest of variables to observe; but accepting this limitation, their almost unique light changes do aid in their recognition and there are relatively large areas of the sky which have so far not been thoroughly examined for the presence of these stars.

REFERENCES

1 Hind, J. R., *M.N.*, **XVI**, 56 (1856)
2 Pogson, N. R., Original Records, Ed. H. H. Turner, *M.N.*, **LXVII**, 119 (1883)
3 Wells, L. D., *H.C.O. Circular*, No. 12 (1896)
4 Peters, C. H. F., *A.N.*, **65**, 55 (1865)
5 Kukarkin, B. V., and Parenago, P. P., *General Catalogue of Variable Stars*, 1st Ed. Moscow (1948)
6 Kraft, R. P., *Ap. J.*, **135**, 408 (1962)

2. The Light Variations of the Dwarf Novae

In the previous chapter we saw how the peculiar light variations of the dwarf novae are so marked that they are relatively easy to distinguish from almost every other class of variable star. Indeed, if it were not for their general faintness they would be among the most conspicuous of all variables. Only the recurrent novae and certain of the short-period stars are in any way similar. In the case of the recurrent novae we find that the interval between successive outbursts is far longer than for any of the U Geminorum or Z Camelopardalis stars with the possible exception of T Leonis.

When we consider the short-period variables, the picture is a little more complex. For instance, some of the Cepheids have periods similar to those of the dwarf novae and although there is no difficulty with the brighter Cepheids (those which can be readily followed throughout the whole of their light cycles), with the fainter members which can be easily observed only when at maximum, some confusion may exist in distinguishing between these stars and the dwarf novae until observations carried out over a period have shown the regularity, or lack of it, of the maxima.

In the present chapter we shall be discussing the light changes of these stars in more detail and in particular, how the study of the variation in brightness has led to the reclassification into two quite distinct groups.

As has been said, the dwarf novae are characterised by sudden and usually steep increases in brightness which occur at intervals of a few weeks or months. The abrupt rise is very reminiscent of the true novae and it is not surprising that the two classes have often been confused at the time of discovery. Where the star is sufficiently bright for the minimum to be observed also, there is usually no such difficulty since the amplitudes of the dwarf novae are smaller than those of the novae, normally lying between two and six magnitudes.

The dwarf novae do, however, share certain other features with the novae such as their characteristic maxima, a binary nature and very high ultra-violet intensities. On the other hand, we find that there are no forbidden lines in their spectra when at maximum, nor is there any spectroscopic evidence of any expanding gaseous shell following the outbursts and it seems quite clear, as we shall see in a later chapter, that the physical processes giving rise to these sudden increases in radiation are not the same in the two classes of star.

There is, though, one quite impressive piece of observational evidence which strongly suggests (although it does not prove), a generic link between the dwarf novae and novae, with the recurrent novae forming an intermediate group between the two. As long ago as 1934, Kukarkin and Parenago[1] discovered that the cycle-amplitude relationship which holds for many of the U Geminorum and Z Camelopardalis variables is the same as that for the recurrent novae and, by extrapolation, possibly for the novae too. Before approaching this question, however, it will be well to consider for a moment what we mean exactly by the cycle, or mean period, of these stars.

Mean Periods of the Dwarf Novae

The induction time between the outbursts of these variables is by no means periodic. On the contrary, even for the more regular members of the U Geminorum class there are wide deviations from the mean cycle length. The individual cycles for U Geminorum, for example, may be anything from 60 to more than 150 days although these figures, particularly the upper limit, must be accepted with reservation since, being a zodiacal star, this variable cannot be observed for about two and a half months each year when the sun is passing through the constellation of Gemini (from about the beginning of June to the middle of August). Undoubtedly several maxima have been missed in this way. When considering SS Cygni, of course, we are on much more certain ground for here we have an almost continuous record going back more than half a century. It is safe to say that our knowledge of the light cycles of SS Cygni is sufficiently complete to allow statistical methods to be applied in an attempt to elucidate the peculiar behaviour of this variable.

Here we find that whereas in most years there are five or six maxima giving an average cycle of about 55 days; in others, notably 1960, there were no fewer than twelve maxima with a thirteenth occurring early in January 1961. Over this particular period of twelve months therefore, the mean period was only 33 days.

Quite obviously, a year is far too short a time over which to define the mean period of any of these variables and the values at present accepted have been obtained by averaging the cycle lengths over as long a period as possible, preferably not less than ten years. Naturally, there are only a few dwarf novae which are bright enough to have been observed sufficiently completely for reliable mean periods to have been determined. One of the problems facing astronomers at the present time is to extend observations of these stars to include many more members of this class. The great majority of these variables are extremely faint and since the large telescopes in the professional observatories are committed to other cosmological problems for which they alone are suited, much of the work of continually following the dwarf novae rests in the hands of amateur observers. It is perhaps inevitable that this has resulted in the brighter variables being closely observed almost to the exclusion of the others. Fortunately, this situation is being remedied and many more of the dwarf novae are being thoroughly observed.

As far as the small subgroup of the Z Camelopardalis variables is concerned there is, of course, the added complication of the 'standstills' to be taken into account making it even more difficult to assign accurate mean periods to these stars.

With these qualifications in mind, the mean cycle lengths of about a third of the known dwarf novae have been determined with more or less precision, most of this work being carried out by members of the various variable star organisations, notably the Variable Star Section of the British Astronomical Association and the American Association of Variable Star Observers who have contributed more than a quarter of a million estimates during the past century.

When we remember that most of these variables have maxima fainter than magnitude 12.0, this is a truly remarkable achievement.

One result which has emerged from this work is that, taken as a whole, the Z Camelopardalis variables have shorter mean periods than the U Geminorum class although as we can see from Fig. 3 there is considerable overlap between the two groups.

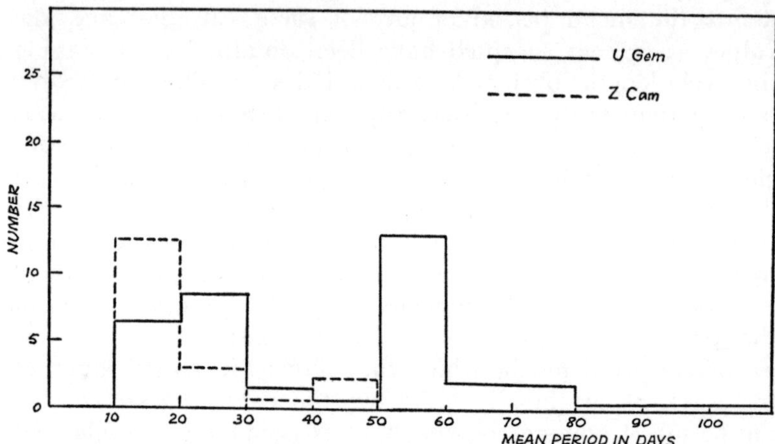

Fig. 3 – Distribution of the mean periods of the U Geminorum and Z Camelopardalis variables.

Whether this fact is significant and indicates a real difference between these two groups is a problem we shall defer until a later chapter when we come to discuss the binary nature of these stars.

Cycle-Amplitude Relationship

The similarity between the light variations of the dwarf novae and the ordinary novae has been frequently pointed out in the past[1] and led to the assumption, which appeared reasonable at the time, that the physical causes underlying the outbursts and related phenomena were fundamentally the same in both classes of variable. We must not lose sight of the fact, however, that at this time very little was known of the spectral changes occurring in these stars, especially when they are at minimum light. It is only comparatively recently that high-dispersion spectrograms taken with the 100- and 200-inch reflectors, have been obtained.

These have provided us with conclusive proof that a fundamental difference exists between the outbursts of the dwarf novae and those of the novae themselves. Since the beginning of the present century, almost every bright nova has been examined spectroscopically shortly after the initial rise to maximum and there is ample evidence for the ejection of a mass of gas, mainly hydrogen, moving outward from the star with high velocities indicative of an explosive condition within the outer atmospheric levels. In the case of the supernovae, the explosion is considerably more violent and much of the mass of the star is ejected at the time of the outburst. The spectra of the dwarf novae at maximum show no indication of this whatever. Clearly the sudden sharp increase in brightness is not brought about by the ejection of mass even on the somewhat lesser scale we might expect considering the smaller amplitudes. Since this anomalous result has a direct bearing on the binary nature of these stars we shall defer discussion on this topic until a later chapter.

The early investigators of these variables faced one further difficulty when they attempted to utilise their observations of the U Geminorum stars to elucidate the various problems posed by the novae as suggested by Payne and Gerasimovic[2]. If there is this similarity between the abrupt maxima of the dwarf novae and the more spectacular outbursts found in the novae, the question which immediately springs to mind is: Why do the former occur fairly frequently while in the novae there is normally only one outburst in several thousand years? The obvious situation, of course, would appear to be that there must be some correlation between the scale of the outburst (as measured by the amplitude and the area under the light curve) and the length of the mean cycle; the tremendous explosions of the novae resulting in a much longer period of relative quiescence.

Is there any observational evidence in support of this? Unfortunately, the problem of establishing such a relationship is by no means as simple or as straight-forward as might appear at first sight. Only a mere handful of novae are known (the recurrent novae such as RS Ophiuchi and T Pyxidis) where more than one outburst has been recorded thereby enabling us to determine some kind of mean cycle for these stars. Certainly it is significant that these recurrent novae have smaller amplitudes (usually less than nine magnitudes) than the typical novae with amplitudes between ten and fourteen magnitudes although since the discovery that most, if not all, of the dwarf and recurrent novae are binaries, it seems quite probable that the true range of the nova companion is greater than that observed visually.

Is the position any better, however, when we examine the dwarf novae? Here too we find that the number which have been extensively examined, and for which reasonably accurate mean cycles can be derived, is also very small. It is true that we have long and continuous observations for SS Cygni and its southern counterpart VW Hydri but in most other cases it is extremely difficult to determine either the amplitudes of the individual maxima or the lengths of the cycles owing to a paucity of observations, or even to their total absence especially around minimum.

We have already seen that U Geminorum itself is a case in point. Not only is this star very faint at minimum but during the late spring and early summer the sun passes through Gemini and the variable is then totally unobservable. This means that several maxima have been missed altogether. A similar unfortunate state of affairs exists for VZ Aquarii, UV Geminorum, X Leonis, FQ Scorpii and the other zodiacal stars. Evidently any attempt to derive a cycle-amplitude relationship must take these considerations into account and it must also be borne in mind that we are dealing with only a small sample of the total number of known dwarf novae. How far this sample is representative of the group as a whole is difficult to assess at the present time.

Nevertheless, as Kukarkin and Parenago[1] have shown such a correlation does exist, which becomes linear (Fig. 4) if the logarithm of the mean period is taken instead of the period itself.

From the slope of the line in Fig. 4, Kukarkin and Parenago obtained the following equation

$$A = 0.63 + 1.667 \log P \qquad (1)$$

where A and P are the mean amplitude and mean period respectively.

It must be noted at this point that only eight stars were considered in this investigation, four U Geminorum variables, two Z Camelopardalis stars and two recurrent novae. Several of those which might otherwise have been included had to be

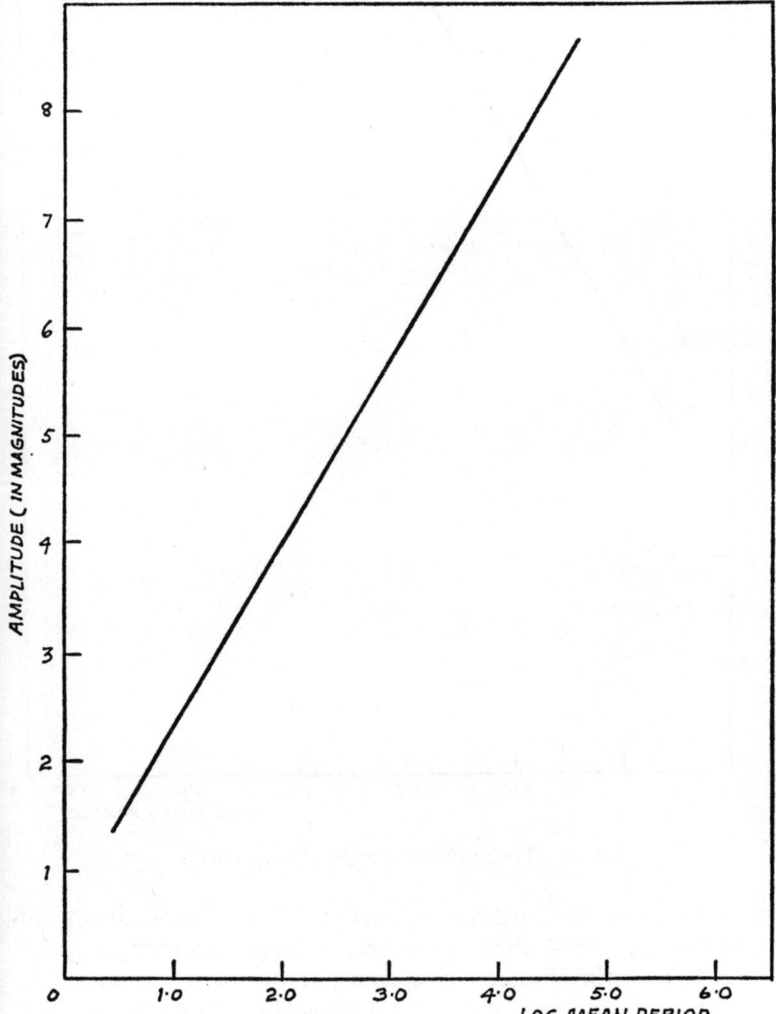

Fig. 4 – The cycle-amplitude relationship (*After Kukarkin and Parenago*).

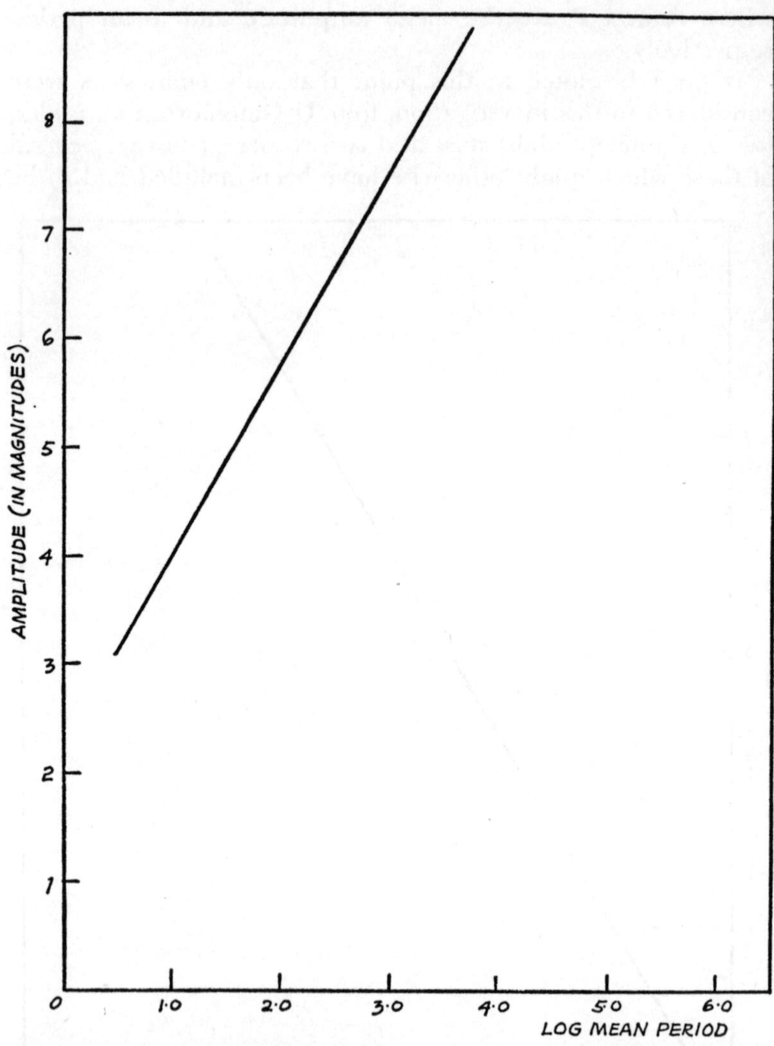

Fig. 5 – The revised cycle-amplitude curve.

rejected from the argument owing to an insufficiency of observations. Since then, however, a large number of additional light estimates, covering many more dwarf novae, have accumulated and Fig. 5 depicts a revised curve drawn by the author from the light curves of fifteen U Geminorum, seven Z

Camelopardalis variables and five recurrent novae. All of the results are on the visual scale since for those stars where both visual and photographic magnitudes are available, it has been found that in almost all cases, the amplitudes based on the photographic scale are greater by about one magnitude.

The revised equation connecting mean cycle and amplitude found from the above curve is

$$A=2.28+1.85 \log P \qquad (2)$$

In general, the scatter of the results about the mean is quite small with the particular exception of RU Pegasi for which the visual range of 2.5 magnitudes is too small for its mean period of 70 days. Unless there is something very peculiar about this particular variable, it would appear that our estimate of either the amplitude or the mean period is wrong. Although not a zodiacal star, RU Pegasi is not readily observable during the winter months when it rises shortly before dawn. In spite of this, however, we have sufficient data to be reasonably certain of the mean period and accordingly we must look to the amplitude for a possible explanation of this curious anomaly. When at minimum, high powers are necessary for the observation of this star since it has a very close companion of roughly the same magnitude from which it must be clearly separated and it is possible that some small error may be introduced here, especially if averted vision is employed in making a light estimate. The fact that the scatter at minimum is usually very small, however, would suggest that we must look for some other explanation and in a later chapter, when discussing the binary nature of these variables, we shall consider a more likely reason for this apparent anomaly.

Extending this interesting point still further, we may ask if it is possible to obtain a similar relationship between the individual cycle lengths and amplitudes for a single variable. If so, how do such results fit in with those we have obtained for a series of these stars? We have seen that if we consider fairly short periods, of about a year or so, the mean cycles thus derived often vary quite considerably from that determined over, say, a quarter of a century. Consequently, we might expect that the mean amplitudes and cycle lengths found during these much shorter periods will show a marked variation.

We must, of course, choose a variable which has been observed over a long period and for which not a single maximum has been missed throughout that time in order to obtain a reasonably significant result. Clearly, the logical choice is, once again, SS Cygni. From a study of the light curve of this star produced from more than 50,000 observations covering 35 years and some 240 cycles, Kukarkin and Parenago[1] have arranged the cycle lengths and the mean amplitudes of the corresponding maxima in order of increasing period into four groups (Table 1).

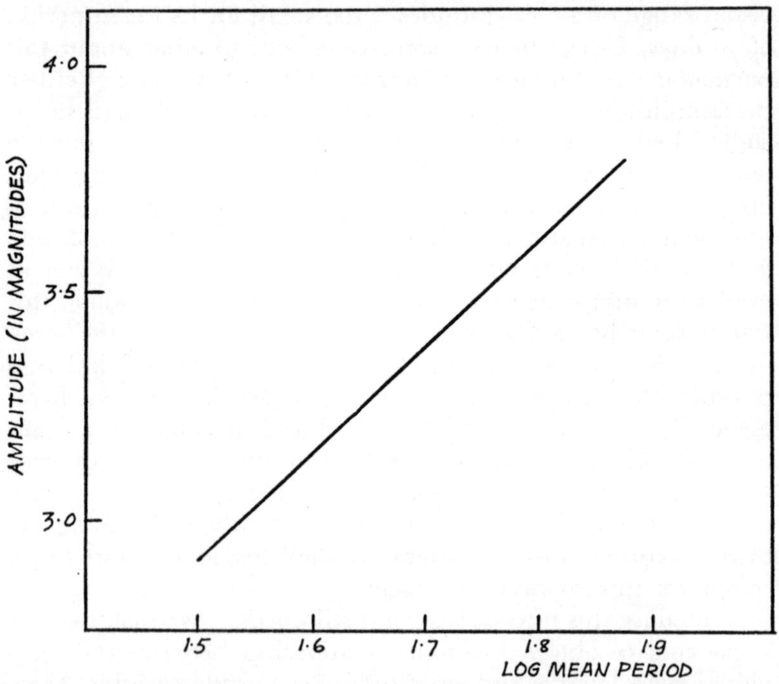

Fig. 6 – Cycle-amplitude curve for SS Cygni (*After Kukarkin and Parenago*).

There is clearly a distinct correlation between period and amplitude for this star and furthermore, it is found that the results also fit well the linear relationship shown in Fig. 4.

Although there is a dispersion of the single points about the straight line this may be ascribed to the inhomogeneity of the

Table 1

CYCLE-AMPLITUDE RELATIONSHIP FOR SS CYGNI

Mean period (days)	Long mean period	Mean amplitude (magnitude)	Number of cycles
33.8	1.53	2.88	60
45.6	1.66	3.29	60
56.4	1.75	3.45	60
71.0	1.85	3.65	57

various photometric scales used by different groups of observers and we must also remember that visual amplitudes do not provide us with a really accurate measure of the intrinsic intensity of these outbursts since these will obviously be influenced to a certain extent by variations in the intensities of different spectral lines.

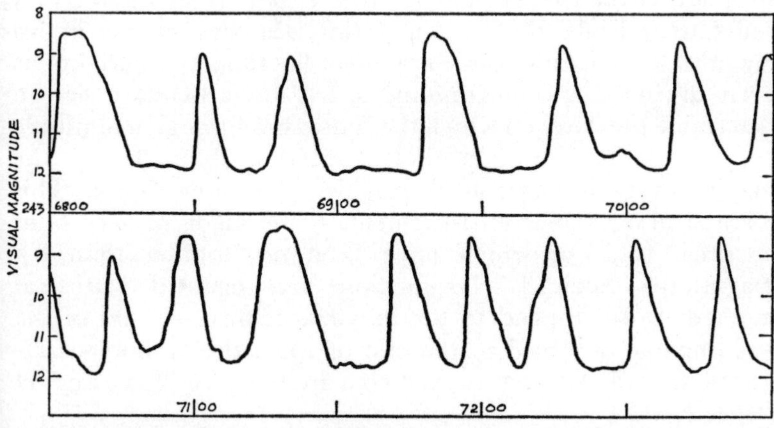

Fig. 7 – Typical light curve of SS Cygni (*Courtesy of the British Astronomical Association*).

Light Curves of the Dwarf Novae

The properties of most of the dwarf novae are very similar to those of U Geminorum itself and it is for this reason that those stars which do not exhibit the added characteristic of the 'standstills' have been designated the U Geminorum variables. Generally speaking, we find that the maxima are of two main types – long and short – depending upon the time spent at maximum, with the former being usually the brighter of the two (Fig. 7).

This does not mean that other types of maxima are not exhibited by certain stars. Let us begin by considering SS Cygni. Very often, this variable undergoes an anomalous type of outburst in which the rise from minimum is protracted, sometimes accompanied by marked fluctuations on both the ascending and descending branches of the light curve. At the same time we usually find that the star attains a brightness about a magnitude below that of a typical long maximum. Since hardly a single maximum of this star has been missed in all of the time since its discovery in 1896, this fact enabled Campbell[3] to make a very comprehensive classification of the maxima of this particular variable. Basing his study on almost 45,000 observations, he has arranged the maxima into four main groups termed A, B, C and D in order of decreasing steepness of the rise. A number from 1 to 10 is also added as a suffix to indicate the length of time the star spends above magnitude 10.0. Since there are minor fluctuations in brightness even during a flat maximum, it has been found easier to determine the dates on which the star passes through magnitude 10.0 on both ascending and descending branches of the maximum than when it is brighter than this figure. This method also enables measurements to be made of very faint maxima, some of which have been no brighter than 9.5 magnitude. Such a classification had enabled statistical methods to be applied to the maxima of this star and of the 504 maxima observed to the end of 1967, the author has assigned 62, 16, 12 and 10 per cent to types A, B, C and D respectively.

Another U Geminorum star which deviates somewhat from the simple scheme of long and short maxima is SU Ursae Majoris. The average cycle of this variable is quite short being only about 16 days. In step with this, we also find that its amplitude is relatively small, varying between 12.0 and 14.5 magnitude with the star remaining bright for only one or two days before fading quite rapidly. About every six months or so, however, a 'supermaximum' occurs (Fig. 8) with the star reaching 10.9 magnitude and remaining above minimum for a period usually between 15 and 25 days.

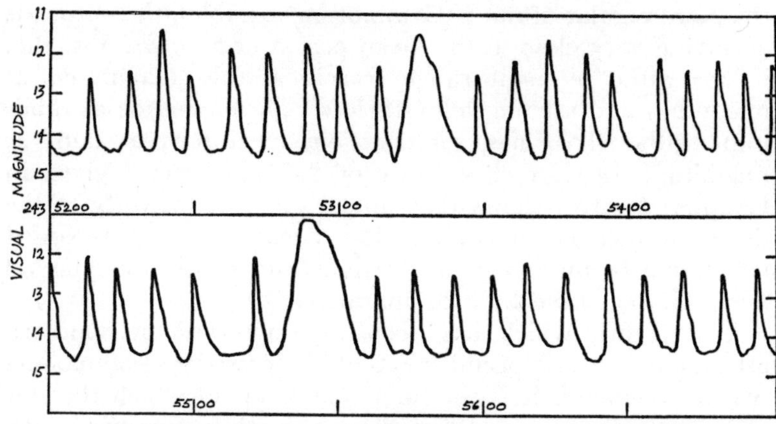

Fig. 8 – Light curve of SU Ursae Majoris (*Courtesy of the British Astronomical Association*).

Over the period during which this star has been under close observation, such 'supermaxima' have happened with a high degree of regularity. Very little theoretical work appears to have been carried out, however, on this phenomenon which, as we shall see in the second part of the book, is not confined to this one star but is shown, in varying degrees, by certain other U Geminorum variables. One may even argue that every U Geminorum star shows such 'supermaxima', identified with the classical long maxima of stars such as U Geminorum and SS Cygni although, in this case, they occur far more frequently than in SU Ursae Majoris. If this is so, then they appear more conspicuous in the latter case merely because of their infrequency and the wide difference between the two types of maxima of this star. If, however, taking into account the marked regularity with which these 'supermaxima' occur, we incline to the other view and regard them as abnormal, then from the light curve it would appear almost as if there is a secondary induction period of some 160 days in this star, over and above the 16-day period of the normal maxima. All of this, of course, must depend to a large extent on the nature of the outbursts of these stars and as this will be discussed in detail in a later chapter, we will defer this problem until then.

Mention must also be made here of RU Pegasi. Observations made over more than 30 years have shown that this is one of the more regular of the U Geminorum variables, the outbursts occurring very close to the mean period of 70 days. Visually, it has quite a small range, reaching 10.0 magnitude at maximum and seldom fading below 12.5 magnitude at minimum although it has, on occasion, been seen as faint as magnitude 13.1. A close study of the light curve (given in Chapter Twelve) shows that unlike the vast majority of the dwarf novae, no maximum has been definitely recorded which can be unambiguously assigned to the abrupt type, all being characterised by a comparatively slow rise.

VW Hydri, which has been well observed by southern astronomers, is also peculiar in that of the two types of maxima which are found, it is the rarer, flat kind, in which the star takes between three and four days to reach maximum, which is the brighter of the two. In the much more frequent type there is the usual steep rise to a sharp peak, followed by a quite rapid decline. The situation here is somewhat similar to that of SU Ursae Majoris (Fig. 8).

We must now return to the point made at the beginning of this chapter, namely that an examination of the light changes of these stars has enabled astronomers to divide the dwarf novae into two distinct subgroups. In spite of the various differences which have just been outlined, the U Geminorum variables form a fairly homogeneous class and in particular their maxima are reasonably smooth, the decline to minimum being regular and not markedly interrupted.

A few years after its discovery, it was noticed that one star, Z Camelopardalis, possessed one distinguishing feature which differentiated it sharply from the other variables then known. After certain maxima, the star did not fade to minimum but remained at some intermediate, almost constant brightness, for a period which could be anything from a few days to more than a year. These 'standstills' as they are called can only be said to end when the star has declined to a normal minimum after which the more usual light variations are then resumed. On at least one occasion, however, Z Camelopardalis[4] has brightened to a maximum immediately following a 'standstill'. At other times, the light variations of this star (and others of the same type) are completely irregular and totally unlike those of the U Geminorum variables.

As we have already seen, very few Z Camelopardalis variables are known but we must not lose sight of the fact that the majority of the dwarf novae are extremely faint objects and are still only intermittently observed. There is consequently always the possibility that further observations may result in the transfer of some of those variables at present classed as U Geminorum stars into this group. Equally obviously, the 'standstills' do not occur with the same frequency in different stars. Some are known in which they are extremely infrequent making the problem of assignment to the Z Camelopardalis group even more difficult.

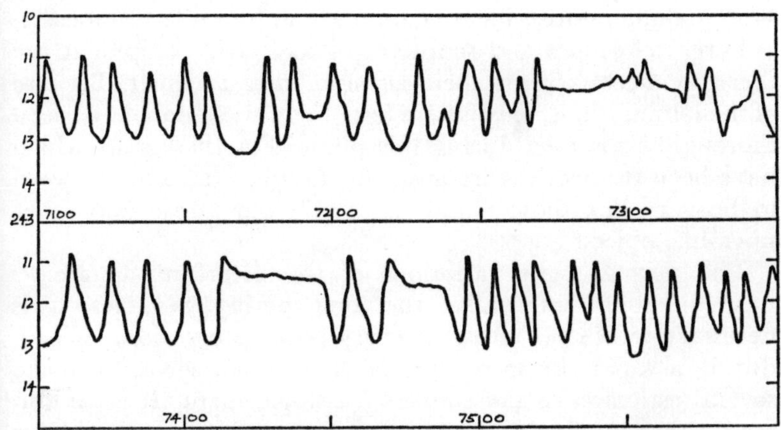

Fig. 9 – A portion of the light curve of RX Andromedae.

Apart from the 'standstills' and occasional periods of erratic behaviour, the two groups of stars are very similar although the amplitudes of the Z Camelopardalis variables are usually smaller, their mean periods shorter and the time spent at minimum less than those of the main class of dwarf novae.

The light curve of a typical Z Camelopardalis variable is shown above in Fig. 9.

Photoelectric Observations

So far we have been considering only those light changes which occur during an outburst. These are the major variations in brightness and for several of these variables are relatively easy to observe with moderate apertures. We must now discuss other minor fluctuations which take place when the star is at minimum; changes which, as we shall see, are in many ways more important when we come to probe into the physical nature of these variables, than the behaviour of the star during maximum.

Here we are immediately faced with several difficulties. At minimum, all of these stars are faint objects, very few being brighter than fourteenth magnitude and the light changes with which we are concerned are both rapid and small, seldom exceeding 0.05 magnitude. Then too, many of these variables lie in crowded fields close to, or within, the Milky Way making positive identification at minimum far from easy.

Large telescopes and sensitive photoelectric equipment are therefore necessary for their accurate measurement. Because of these limitations, only a bare handful of these stars have been thoroughly observed during this phase. For those stars which have been studied, the irregular fluctuations have been shown to be very like those found in certain old novae and other nova-like objects[5].

The normal light variations of the dwarf novae are so unpredictable that, unlike the long period variables, it is necessary to observe them on every possible occasion; indeed, during a rapid rise to maximum, it is advantageous to make several estimates in the course of a single night. It is for this reason that by far the greater proportion of observations have been made visually, mainly by the members of the various amateur organisations. In spite of the personal errors which are inherent in the making of visual estimates, this method has proved to be quite suitable for the production of light curves for such stars.

When we come to the study of the small, irregular fluctuations exhibited by these variables, especially around minimum, such visual estimates are quite inadequate. Since they amount to only a few hundredths of a magnitude at most it is necessary to use a far more accurate method which is not subject to human error. Such a method is found in the photoelectric cell. Although the basic principles of the photoelectric effect properly belong to the realm of physics, the astronomer has made so much use of it that a brief description will not be out of place here. The photoelectric effect consists of the emission of electrons from a thin film of a metal such as calcium when light falls upon it provided that the energy of the incident light is sufficient to eject the outermost electron, or electrons, from their orbits within the electron cloud of the metal atom.

Briefly, a photoelectric cell consists of an evacuated quartz or glass bulb whose inner surface has been coated with a thin deposit of an alkali, or alkaline earth, metal – usually sodium, potassium or caesium although as mentioned above, calcium may also be used – this forming the cathode, its function being to emit electrons on irradiation by light of a suitable wavelength. The anode is a wire or grid mounted in front of the cathode to collect these emitted electrons. In order to facilitate the passage of the electrons between cathode and anode, a potential difference of the order of 100 volts is applied to the electrodes. The resultant charge on the anode is measured by means of an extremely sensitive galvanometer after amplification, normally by the use of a refrigerated photomultiplier.

Within certain limits, the photoelectric current generated is directly proportional to the intensity of the incident light and where we are dealing with the measurement of very small and rapid light fluctuations, the photoelectric cell has one very great advantage over the photographic plate, namely that the time necessary for a single determination is extremely short – a readout time of as little as five seconds being quite feasible with modern techniques. This enables astronomers to take readings in ultra-violet, blue and visual light in rapid succession merely by changing the requisite filters. Similar determinations by the photographic method would require several hours as there are ancillary operations which have to be carried out and such a time lapse would be out of the question where such rapid and irregular light variations have to be closely followed.

Photoelectric observations have been used to great effect in the examination of the dwarf novae by several astronomers but here we must anticipate things a little and mention something of the spectra of these stars. The first comprehensive survey of these variables and other nova-like objects was carried out in 1943 by Elvey and Babcock[6] at the McDonald Observatory. For most of these stars, the spectra at minimum were found to consist of broad emission lines of hydrogen, fainter lines of helium and the bright H and K lines of ionised calcium. These are superimposed upon a continuum which, according to the distribution of intensity, corresponds to a spectral type of G or K. At maximum, the spectra appear to be of a much earlier type, either B or A, with shallow absorption lines of hydrogen and occasionally narrow emission lines which are always faint. Owing to the faintness of these variables, only low-dispersion spectrograms were obtainable by Elvey and Babcock using exposures of between two and three hours; conditions wholly unsuited to the resolution of any fine structure of the emission lines.

Even earlier, Joy[7] had shown that the spectrum of RU Pegasi at minimum possesses, as well as the usual emission lines, the absorption lines corresponding to a type dG3 star. Sixteen years later[8], he was also able to show that SS Cygni has a composite spectrum at minimum, being a spectroscopic binary with an orbital period of 6 hours and 38 minutes, the component stars having spectral types of sdBe and dG5.

More recently, Kraft[9] found U Geminorum itself to be a single-lined spectroscopic binary with a period of 4 hours and 10 minutes and a spectrum at minimum which consists of broad Balmer lines in emission which are clearly double.

We are now in a position to appreciate the importance of the photoelectric observations of these variables, especially when at minimum, since the fact that the emission lines are double (in the case of U Geminorum at least), suggests that here we may find some evidence of eclipse. The peculiar spectra indicate that most, if not all, of these stars are binaries and as their orbital planes will be randomly oriented in space, a fair proportion of them will have the planes of their orbits aligned with the line of sight as seen from Earth and it should, theoretically, be possible to detect eclipses as the stars revolve around their common centre of gravity.

Now we come to another crucial point. Since all of these stars are subject to major increases in brightness at completely unpredictable intervals, which is the optimum phase during which to look for evidence of eclipse, bearing in mind that any diminution in light will be very small? Quite obviously it will be difficult, if not impossible, to detect such minute light

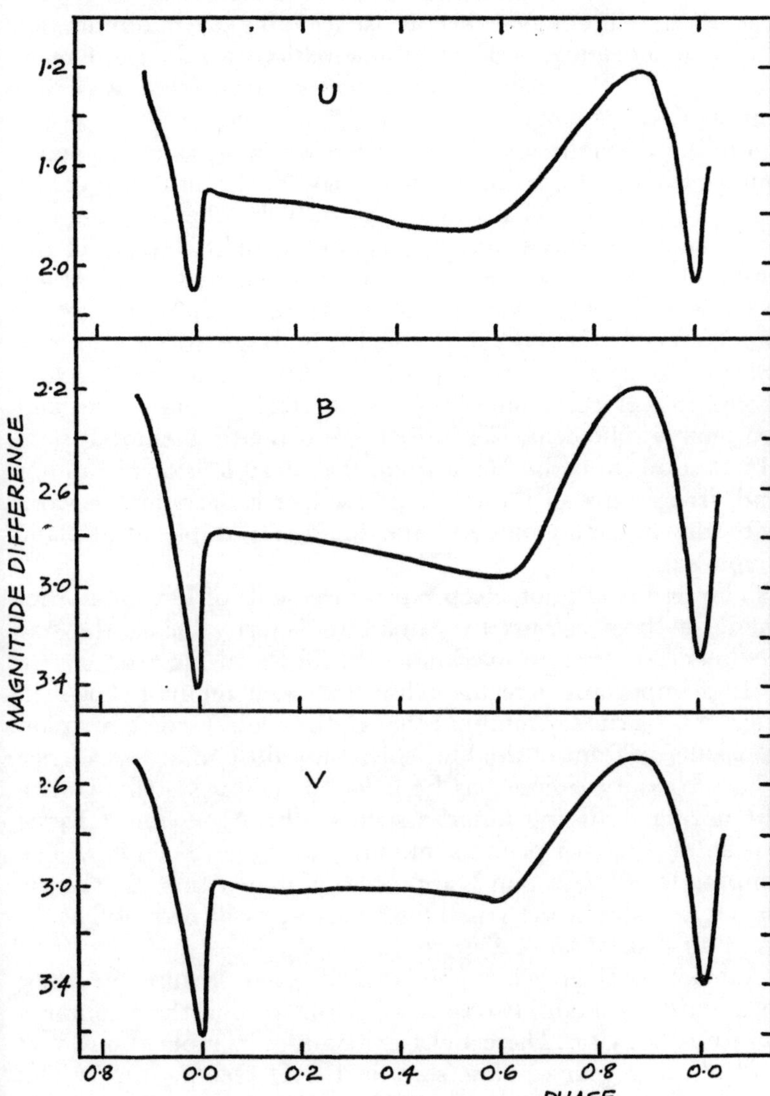

Fig. 10 – Photoelectric light curve of U Geminorum made by three-colour photometry (*Astrophysical Journal by kind permission of the University of Chicago Press*).

changes during the times when the star is increasing in brightness to maximum or when it is declining to minimum for under such conditions the small fluctuations will be completely obscured by the major light variations associated with the outburst.

Similar difficulties are encountered when the star is at maximum although from the point of view of the intensity of the incident radiation this would naturally be the best time to make such measurements. At maximum, light changes of the order of 0.3 magnitude are quite common and in addition, leaving aside the actual physical nature of the outburst for the moment, the effect is sufficiently large to swamp any minute changes due to a possible eclipse. We are therefore forced to seek this evidence when the star is at minimum and reasonably quiescent. Certainly there is the problem that such observations must be made when the variable is at its faintest and, irrespective of the telescope used, it is necessary to work with the highest gains and the smallest possible focal-plane diaphragm.

The results of photoelectric measurements of U Geminorum made in three colours by Mumford[10] shortly before the star commenced a rise to maximum are shown in Fig. 10.

It is important here to realise that both minima shown in Fig. 10 are primary minima; that is, the cooler, red companion is passing in front of the blue star. Very little, if any, evidence of a secondary eclipse has been found in this system. Unlike the normal eclipsing binaries such as the Algol and β Lyrae variables, the two primary minima are quite dissimilar. The amplitude of the second is appreciably smaller than that of the first, especially in the visual light curve as will be readily seen from an inspection of Fig. 10.

Obviously then, we cannot explain these features in terms of a simple system of two stars revolving around their common centre of gravity. There must exist added complications over and above a pair of close stars in the U Geminorum system and as yet several of these features have not been fully elucidated. Perhaps we may be able to visualise the differences which exist between the two eclipses a little more easily from the composite light curves shown in Fig. 11.

On the basis of these curves, it would appear that the visual minimum is the most altered. Not only is the second minimum changed in shape but it is also filled in relative to the first. Unfortunately, the observations do not tell us which of the two eclipses is abnormal.

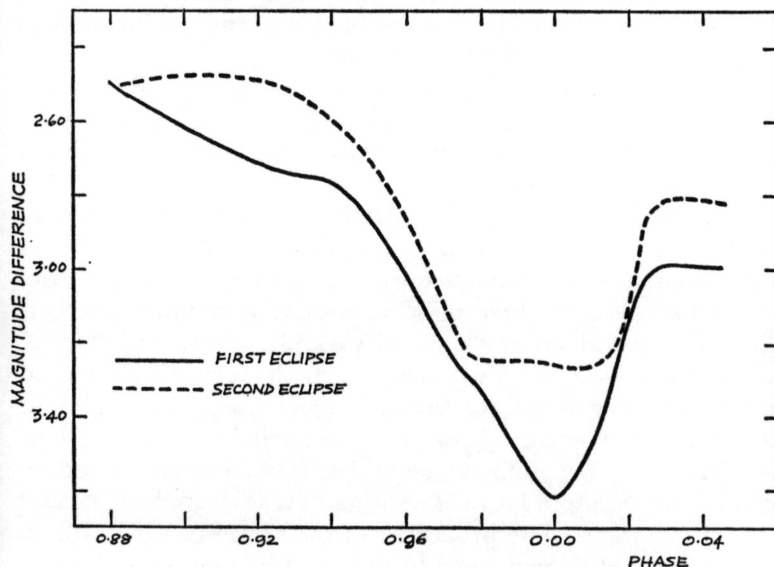

Fig. 11 – Composite light curve of the U Geminorum eclipse from visual estimates (*Astrophysical Journal by kind permission of the University of Chicago Press*).

One important question which remains to be answered is whether the eclipse is total, partial or annular. Clearly this depends upon several factors; the inclination of the orbit, the relative dimensions of the two components and whether the blue object in the system is a single entity or surrounded by a ring or disc of gas. How we determine the values of these parameters is a problem we shall take up at a later stage. At present it will be sufficient to say that the orbital inclination of U Geminorum is very close to 90° and that the cool component is a late-type dwarf star which, since it fills its lobe of the zero-potential surface, is larger in radius than the blue companion. As for the last possibility, there is direct observational evidence that such a ring of gas does encircle the blue star.

From the spectroscopic data obtained by Kraft[9], it is clear that when the blue star is at superior conjunction (at the time of a primary eclipse), the emission spectrum is still visible although diminished in intensity. Now this could indicate an annular eclipse but it is more likely that such a weakening results from a total eclipse of the blue star and an incomplete eclipse of the surrounding ring of gas.

Other arguments which favour a total eclipse of the blue component are the reddening of the system at minimum and the fact that from Fig. 10 we see that the blue light is diminished relative to the ultra-violet.

The situation is obviously extremely complex and here we shall digress a little to consider other nova-like variables in order to see if a more detailed understanding may be got from an examination of their light variations at minimum. Now there are several small groups of variable stars which, like U Geminorum, are eclipsing systems with very short orbital periods. Among these are the old novae which, at minimum after the outburst, also show very similar intrinsic fluctuations as well as a pronounced shoulder immediately preceding eclipse, for example Nova T Aurigae (1891)[11] and Nova DQ Herculis (1934)[12]. The properties of the novae before outburst are virtually unknown since in only one or two isolated cases do we know anything about the pre-nova stage of these stars.

There is also a small class of very peculiar variables which, although there is as yet no direct proof that they are also old novae, again show similar fluctuations in light. This group is typified by stars such as UX Ursae Majoris and RW Trianguli. The former has been extensively studied by several astronomers[13,14,15] and the latter particularly by Walker[16]. Much of the available evidence, both photoelectric and spectroscopic, indicates that these variables are close binaries consisting of a white dwarf and a much larger red companion. We shall go into this in more detail in a later chapter. At present, it is sufficient to state that the radius of the red star is such that it has already filled its lobe of the zero-potential surface and gaseous material is overflowing through the inner Lagrangian point and taking up an orbit around the white dwarf companion, a situation believed to be very similar to that pertaining in the U Geminorum system.

Finally, mention must be made of one further class of variable which many astronomers believe may be the precursors of the dwarf novae. These are the W Ursae Majoris variables, also known as the contact binaries. Relatively few of these particular variables are known at present, this being almost certainly due, not to their comparative rarity within the galaxy (they are actually believed to exist in fairly large numbers), but to their intrinsic faintness. Their light curves show primary and secondary minima of almost equal depth and a continuous light variation similar to the β Lyrae stars, the interpretation being that here we have two stars of almost equal luminosity which are grossly distorted by tidal and gravitational forces due to their proximity. There is also spectroscopic evidence of a gas stream in these systems flowing from the primary to the secondary, the former evolving more rapidly than the secondary, presumably towards the white dwarf stage.

AE Aquarii

At this point we must consider the light variations of one particular U Geminorum variable which is so atypical of the

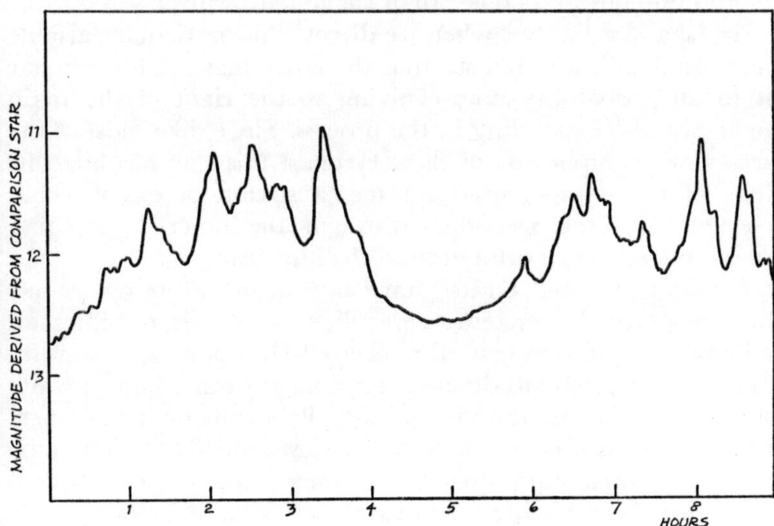

Fig. 12 – Light curve of AE Aquarii obtained from photoelectric measurements.

class that it deserves special mention. This is the star AE Aquarii whose light curve (Fig. 12) is totally unlike that of any other dwarf nova. Here we find very rapid increases of brightness which have an average amplitude of only about half a magnitude, the fluctuations being of an extremely complex nature.

From Fig. 12, it will be seen that there is very little periodicity about these small-scale outbursts although it is possible that there exists an underlying period of about one hour. More violent outbursts of up to two magnitudes have been observed by Zinner[17] but these are very rare occurrences.

In 1940, Joy[7] discovered that AE Aquarii is a spectroscopic binary with a period of about 17 hours and classified the component stars as of spectral types sdBe and K0, the latter having a mass similar to that of the sun. Sixteen years later, Crawford and Kraft[18] used the Lick 36-inch refractor to obtain high-dispersion spectrograms of this star in order to make an accurate classification of the late-type component, comparing it with two standard K-type dwarf stars, HD 109011 and HD 154363. From this investigation, they concluded that the secondary is a K5 star of luminosity class IV or V, somewhat more luminous and cooler than suggested by Joy.

In Chapter Twelve, when we discuss this particular variable in more detail, we shall see that the fairly massive K-type star is in all probability now evolving to the right of the main sequence and expanding in the process. Since, like most of the secondary components of these systems, this star has filled its lobe of the equipotential surface, a stream of gas is being ejected from the secondary through the inner Lagrangian point, taking up an orbit around the blue primary.

As Crawford and Kraft[18] have shown, the white dwarf star in this particular system is peculiar in that it occupies an anomalous position on the Russell-Hertzsprung diagram, being about 4 magnitudes brighter than any other known white dwarfs, yet lying some 10 magnitudes below the main sequence. One very plausible explanation of this anomaly is that some of the gas surrounding this star is being continuously collected by it, the amount being estimated as 10^{22} kilograms every year. The probable evolution of systems such as AE Aquarii and the other dwarf novae will be discussed in a later chapter. Here we may simply note that much of the available data suggests that the white dwarf companion was initially the more massive of the two and consequently was the first to evolve, expanding to fill its lobe of the equipotential surface and then losing hydrogen-rich matter through the inner Lagrangian point to the secondary. Once much of the hydrogen was lost in this way, the star contracted fairly rapidly to its present white dwarf state and the secondary then began to evolve, the mass loss reversing direction.

As we have seen, a very odd feature of this variable is the form of the light curve which is totally unlike those of the other dwarf novae, although in other respects it bears a close resemblance to them. One suggestion which has been put forward to explain these relatively minor outbursts of light is that, unlike the other members of this class, small-scale explosions occur in this system, possibly at some distance from the white dwarf star.

Statistical Analysis of the Light Variations

Several astronomers have attempted to apply the methods of statistical analysis to the problem of the light changes of the dwarf novae, in particular to SS Cygni which, as we have seen, is the most comprehensively observed of all these stars.

Lortet-Zuckermann[19] has examined the sequence of maxima of SS Cygni for the period 1896 to 1957, a total of 444 maxima in all, in order to determine whether the various types of outbursts may be considered as a simple Markov chain as originally suggested by Martel[20]. Three categories of maxima have been considered; long, short and weak and to a first approximation it would seem that that the sequence of long and short maxima may be represented by a simple Markov chain. This is probably a result of the undoubted tendency of these two types to alternate.

In the case of the very weak outbursts of this variable, the sequence is fairly well represented by a second order Markov chain and the same appears to be true of the full sequence of long, short, weak and very weak outbursts. However, there are difficulties when we have to postulate Markov chains of the first or higher orders since these do not explain the periods of irregular behaviour when a large number of small outbursts

and relatively few major maxima occur, nor the very large number of cycles or major outbursts. As suggested by Lortet-Zuckermann[19], a recurrent event process may provide a more realistic scheme to explain the full sequence of these maxima.

A full discussion of Markov chains is beyond the scope of this book, dealing as it does with the application of the probability theory. Those readers wishing to pursue the theory further are referred to the excellent accounts given by Doeblin[21,22].

REFERENCES

1 Parenago, P. P., and Kukarkin, B. V., Variable Star Bulletin, *IV*, No. 8, 44 (1934)
2 Payne, C. H., and Gerasimovic, B. P., *A.N.*, **250**, 45
3 Campbell, L., *H.C.O. Annals*, **90**, No. 3 (1934)
4 Holborn, F. M., *J. Brit. astron Ass.*, **70**, No. 3, 134 (1960)
5 Walker, M. F., I.A.U. Symposium No. 3 (1957), Ed. G. H. Herbig (Cambridge, Cambridge University Press)
6 Elvey, C. T., and Babcock, H. W., *Ap. J.*, **97**, 412 (1943)
7 Joy, A. H., *Publ. A.S.P.*, **52**, 324 (1940)
8 Joy, A. H., *Ap. J.*, **124**, 137 (1956)
9 Kraft, R. P., *ibid*, **135**, 408 (1962)
10 Mumford, G. S., *ibid*, **139**, 476 (1964)
11 Walker, M. F., *ibid*, **138**, 313 (1963)
12 Walker, M. F., *ibid*, **123**, 68 (1956); *ibid*, **127**, 319 (1958)
13 Johnson, H. L., Perkins, B., and Hiltner, W. A., *Ap. J. Supplement* **1**, 91, No. 4 (1954)
14 Walker, M. F., and Herbig, G. H., *Ap. J.*, **120**, 337 (1954)
15 Krzeminski, W., and Walker, M. F., *ibid*, **138**, 146 (1963)
16 Walker, M. F., *ibid*, **137**, 485 (1963)
17 Zinner, E., *A.N.*, **265**, 345 (1938)
18 Crawford, J. A., and Kraft, R. P., *Ap. J.*, **123**, 44 (1956)
19 Lortet-Zuckermann, M. C., *Ann. Astrophysics*, **29**, 205 (1966)
20 Martel, M., *These Ann. d'Ast.*, **24**, 267 (1967)
21 Doeblin, W., *Comptes Rendus*, **203**, 24 (1936)
22 Doeblin, W., *ibid*, **205**, 7 (1937)

3. Spectra of the Dwarf Novae

Since the beginning of the century when the increasing number of dwarf novae being discovered provided astronomers with a new and distinct class of variable star whose light variations were markedly different from any other type at that time, there has been much speculation on the nature of these peculiar stars. Apart from the general problem of explaining how such stars can undergo apparently explosive outbursts at frequent, although completely irregular, intervals, other more specific difficulties were soon encountered. The discovery of the 'standstills' among the Z Camelopardalis subgroup provided a problem which is still far from being satisfactorily solved, a situation which is also true of the periods of erratic behaviour in these same variables.

Almost inevitably, there was a tendency to associate these stars with the novae, several of which had been discovered and investigated at this time. It was soon apparent, however, that little further information regarding the physical nature of the dwarf novae would be obtained merely from a study of their light curves.

Digressing a little at this point, we may note that in the whole field of variable star research it is only in the case of the Algol and β Lyrae eclipsing variables that the shape of the light curve alone has provided some explanation of the basic cause of the star's variability before more recent, refined techniques afforded further confirmatory proof. The binary nature of Algol, for example, was suggested by Goodricke long before it was confirmed spectroscopically. In the case of all other classes of variable star we rely very heavily on the spectrum to furnish us with the additional information necessary to allow a model of the variable to be put forward, whether it be of the pulsating type (the Cepheids, long period and semi-regular variables) or of the violently eruptive class (the novae, recurrent novae and supernovae).

Astronomical spectroscopy did not develop fully until the introduction of photography during the closing decades of the nineteenth century, a development which made it possible, not only to obtain a permanent record of these spectra but also to measure the wavelengths of the absorption and emission lines with a high degree of accuracy. It was not until 1872, seventeen years after the discovery of U Geminorum, that the first photograph of a stellar spectrum, that of Vega, was obtained by Draper. So rapid was the development after this, however, that by 1918, the Henry Draper Catalogue which classifies the spectral types of some 225,000 stars was published.

Before discussing the spectra of the dwarf novae, it will be worthwhile here to mention the three instruments which the astronomer has at his disposal for the recording of stellar spectra, namely the objective prism, the slitless spectrograph and the normal spectrograph which employs a slit. Each of these instruments has its own particular use in stellar spectroscopy.

The first is adapted to the photography of large numbers of stellar spectra and is the method used in the programme which culminated in the Henry Draper Catalogue (as we shall see, it would have been an almost insuperable task to produce this tremendous publication by either of the other two methods). The objective prism instrument consists merely of a large prism having a small apex angle which is placed over the objective of the telescope producing the spectra of all stars in the field of view which are of sufficient brightness to affect the photographic plate. It is not necessary to use a collimator with this instrument since all of the stars may be regarded as being point sources of light at infinity and consequently their light rays will be essentially parallel. The stars, however, are still merely points of light and the spectra will be in the form of small points too and in order to widen them sufficiently for detail to be recognised, the image is allowed to trail across the plate by means of the slow motions on the equatorial mounting. Naturally, this method will only provide low-dispersion spectra and cannot be used satisfactorily for the very faint stars.

The slitless spectrograph, on the other hand, although again recording only low-dispersion spectra, may be used in conjunction with very large reflectors and since it is employed at the eyepiece of the instrument, it may be used for photographing the spectra of extremely faint objects, far beyond the reach of the objective prism.

When spectra of a high dispersion are required, we must use the slit spectrograph which is well suited for the study of the relatively bright stars. In this instrument, the light from the telescope is focussed upon the slit of a collimator and then onto a prism or grating before passing into a photographic chamber where the high-dispersion image of the spectrum is formed upon a photographic plate. Such an instrument is of particular value when we wish to measure Doppler shifts which provide both proof of orbital motion in binary systems and an accurate measure of the velocity of motion of each component in its orbit. It is this particular type of spectrograph which has been extensively used in the study of the dwarf novae.

As a class, these variables are not particularly numerous and as we have seen, the majority are very faint objects even when at maximum brightness. It is not surprising therefore that very few high-dispersion spectrograms have so far been obtained although a great deal of work is now being carried out to extend the number of such spectra which, in many cases, have so far been obtained only around the time of maximum. We can readily see how vitally important this branch of astronomical research is, since it is from their spectra that much of our knowledge of the physical nature of the dwarf novae has been obtained. Even with the 200-inch reflector at Mount Palomar, however, the task of gathering this spectroscopic data is both slow and difficult.

The first comprehensive survey of the spectra of the dwarf novae and certain other stars which show nova-like characteristics were made as recently as 1943 by Elvey and Babcock[1] at the McDonald Observatory. If we leave aside AE Aquarii which, as we saw in the previous chapter, is far from being a typical member of this class, only two of the dwarf novae – SS Cygni and RU Pegasi – have normal minima brighter than thirteenth magnitude and for this reason only low-dispersion spectrograms with a spread of about 340 Å per millimetre were obtained during this particular survey when these stars were observed around minimum phase. A higher dispersion was possible only when the variables were undergoing an outburst.

Now this immediately introduces a problem, for such low-dispersion spectrograms of faint stars normally require rather long exposure times of between two and three hours. This is of negligible importance when we are considering stars in which any change is relatively slow, such as the long period variables but such long exposure times are too long to yield sufficient resolution in time for any great detail to be obtained. As a result, it is not to be expected that any evidence of orbital motion would be provided by the spectra obtained by Elvey and Babcock. As we shall see later, this came only when high-dispersion spectrograms were obtained using the largest instruments available.

In the meantime, let us examine in some detail the results found during this initial survey of these variables. At minimum light, the spectra generally consist of very broad, bright lines of the Balmer series of hydrogen, together with bright H and K lines of singly ionised calcium and much weaker emission lines of neutral helium superimposed upon a continuous background. There are, of course, differences from star to star. The intensities of the various emission lines vary appreciably and in some dwarf novae, for example, X Leonis, no emission lines are visible at all.

The underlying continuum also provides us with some important information for it is possible from a study of the continuous background of the spectrum to obtain a good idea of the spectral class of the star producing it. In order to understand why this should be so, we must first consider the region in the star which produces the continuum. This is the zone known as the photosphere which lies just beneath the reversing layer and is a region in which, like the stellar atmosphere, there is a density gradient, the density increasing as we go deeper into the photosphere. Now we find that one property of the photosphere is that whenever light of any wavelength enters this zone it will eventually be totally absorbed and in this respect we may look upon the photosphere of a star as behaving like a black body. If we therefore compare the distribution of energy at different wavelengths throughout the continuum with that of stars of known spectral classes, we may deduce the spectral type of a dwarf novae at various phases of its light variation. For those stars investigated by Elvey and Babcock, the spectral class at minimum, found by this method, was either G or K.

The spectra of these stars around maximum brightness are quite different. Now the energy distribution in the continuum corresponds to a much earlier spectral class, B or A, and the lines due to hydrogen appear in absorption, being usually broad and very shallow. Sometimes, there are also faint and narrow emission features visible.

We now come to an extremely important observation which was made some three years before Elvey and Babcock carried out their monumental work and which proved to be a long-awaited breakthrough in our understanding of these peculiar stars. In 1940, Joy[2] found the absorption lines of a type dG3 star in the spectrum of RU Pegasi when at minimum as well as the usual emission features. This immediately suggests that the variable is possibly a binary system since it appears unlikely that a single star, unless of an exceptionally unusual kind, will possess such a complex spectrum with such wide variations in the intensity distribution. Now it is quite true that these variables, particularly in their very characteristic light variations, are reminiscent of the novae and it may be argued that since the output of radiant energy (mostly in the form of light and heat) in these latter stars is extremely high, the spectroscopic changes recorded by Joy can be explained on the basis of a single star undergoing violently explosive ejections of mass at irregular intervals. This idea is undoubtedly attractive in many ways. Why, therefore, is it no longer accepted by astronomers?

The answer lies in the spectra of these two classes of star. If we examine a typical nova spectrum we find that the emission lines of hydrogen are displaced towards the violet end by an amount which corresponds to a velocity of approach of between 800 and 1200 kilometres per second due to a rapidly expanding mass of gas which is ejected from the surface layers of the star during the nova explosion. Confirmation of this is often provided by visual and photographic examination of the nova some time after the outburst when this small, nebulous cloud of incandescent gas may be seen.

The spectrum of a typical dwarf nova when at, or just

after, maximum, shows no evidence at all of any expanding mass of hydrogen gas, nor are there any of the forbidden lines present which are characteristic of a nova spectrum. We shall defer a full discussion of the probable nature of the outbursts of these variables until a later chapter, merely emphasising at this point that, whatever the cause, it is almost certainly not an ejection of mass similar to that found in the case of the novae.

The emission line spectra of these two classes of star also differ appreciably in two other important respects as has been pointed out by Kraft[3]. Firstly, the excitation of the emission lines in the dwarf novae is much lower than that found in the spectra of the old novae at minimum. The emission line of ionised helium at λ4686, for example, is very rarely observed in the dwarf novae and even when it is present, as in U Geminorum itself, it is extremely faint. Almost all of the old novae, on the other hand, show it in emission, often very strongly. This would suggest that the blue components in the old novae systems have a much higher surface temperature than those of the dwarf novae. Secondly, the emission line spectra of the dwarf nova class are all very similar with only a few exceptions. X Leonis, for instance, shows no emission lines at minimum and Elvey and Babcock[1] have described the spectrum of AY Lyrae at minimum as G-type with no visible emission lines although this has not been confirmed by the more recent work of Kraft[3]. In the case of this particular variable, however, not only is it extremely faint at minimum (photographic magnitude 16.1), but it lies in a very densely populated field of faint stars and it is possible that the variable may have been mis-identified. The old novae show very marked differences among their spectra which have been fully described by Kraft[3] and Greenstein[4]. Here we may mention Nova HR Lyrae (1919) which has an almost continuous spectrum; Nova GK Persei (1901) and Nova DQ Herculis (1934) which both show very strong emission lines, and Nova DI Lacertae (1910) whose spectrum shows features intermediate between the two above extremes with wide, shallow absorption lines upon which are superimposed very narrow, central emission components.

Even now we have not exhausted the information which may

be derived from the spectra of these stars. Both the old novae and the dwarf novae show a curious relationship between the relative luminosity of the secondary component and the orbital period. This is fairly readily observable in their spectra since we find that those variables, of both classes, which have long orbital periods exhibit the spectra of both components. Among the novae this is true for stars such as Nova GK Persei (1901) and the recurrent nova T Coronae Borealis (1866 and 1946) while among the dwarf novae we find a similar state of affairs in the case of Z Camelopardalis, SS Cygni and RU Pegasi. It may, of course, be argued that in those systems where the orbital velocity is high (those with short orbital periods), the increased rotation obliterates the absorption lines. This can be readily shown to be unlikely where there is synchronism of the orbital period and the axial rotation of the secondary and where the cooler star fills its lobe of the equipotential surface. Both of these conditions are thought to hold for these two classes of star and it is now generally accepted that the effect represents a real diminution in the luminosity of the secondary relative to the primary with a decrease in the orbital period. We shall have more to say on this important question when we discuss the binary nature of these variables in a later chapter.

So far, we have discussed the spectra of the dwarf novae only when at maximum or minimum brightness. What is also of great importance, of course, is how the various spectral features change throughout the light cycle, since these provide us with several parameters which are otherwise difficult to obtain. Naturally, the sequence of changes will be easily most followed where the variable is fairly bright at all phases of its light variation and for this reason, the most likely candidate for a spectroscopic survey of this kind is SS Cygni. Not only can the light changes be followed easily but it is well within the range of spectroscopic examination, even when at minimum. Several studies of this star have been made and these will now be discussed in detail.

Cyclic Spectral Variations in SS Cygni

Although very crude, low-dispersion spectra of U Geminorum had been obtained as long ago as 1882 by Copeland[5] and

Knott[6] using an ocular spectroscope, these showed little more than a few emission lines which served to show a link between the star and the few novae which had previously been examined. Some thirty years later, in 1912, Miss Fleming obtained objective-prism spectra of U Geminorum and SS Cygni[7] but again these had a very low dispersion which made it impossible to resolve the many puzzling features shown in these spectra.

Adams and Joy[8] recorded not only the familiar emission lines at minimum and broad absorption features at maximum, but also changes in the relative intensities of the lines throughout the light cycle, a result which was later confirmed by Wachmann[9] who was also able to show that when SS Cygni was at minimum, the spectrum contained bright emission lines of hydrogen and neutral helium 20 Å wide, weak absorption lines of the same width being present whenever the star passed through a long, flat maximum. In addition, although the underlying continuum resembles an A-type star at minimum, this is certainly not the case at maximum brightness.

The first really comprehensive study of the spectral changes occurring in SS Cygni, however, was carried out by Hinderer[10] who made a whole series of spectroscopic determinations covering a long, flat maximum of this variable. These provide a great deal of important information concerning such parameters as the absolute gradient of the spectral line contours, the colour and radiation temperatures and the continuous variation of spectral class throughout a typical maximum.

In general, the development of the spectrum may be divided into three reasonably distinct phases. At maximum, only absorption lines are visible with the underlying continuum corresponding to an A1 or A2-type star. As the variable begins to fade, entering upon the second phase recognised by Hinderer, the spectrum becomes composite with a central emission line appearing which gradually gives way to a double absorption line as the brightness of the star approaches magnitude 10.0. As the decline continues, the spectrum has the appearance of a series of bright emission lines which are flanked by fairly weak absorption features that slowly fade while the emission lines become stronger towards minimum.

At the same time that this is happening, the underlying continuum is gradually weakening with the peak intensity shifted towards the red, providing an explanation of the colour changes of this variable, changes which were noticed as long ago as 1932. The spectral type estimated by Hinderer, reaches F0 at magnitude 10.6 and then passes through the various subdivisions of this type, becoming G1 at 11.6 magnitude and type G4 by the time the star has attained minimum around 11.8 magnitude. At this phase, there are very strong emission lines of both hydrogen and neutral helium, together with the K line of singly ionised calcium.

The radiation temperature varies gradually from 9,600°K at maximum to 5,650°K at minimum, the effective radius being 1.79 times greater at maximum than at minimum. The Balmer jump appears in emission at the limit of the photographic range in the violet. During the rise from minimum (which is, of course, usually more rapid than the decline), the above spectral changes are reversed. When the colour temperatures are measured from the gradients of the spectral lines, they indicate a somewhat wider range than the radiation temperatures, going from 5,000°K to about 13,400°K from minimum to maximum.

From the results obtained during this spectroscopic examination, Hinderer calculated a probable absolute magnitude at minimum of +4.5, corresponding to a parallax of only 0".004 which is appreciably smaller than that found by Kraft[11] which yields a value of +7.5 for the absolute magnitude. Since this latter figure has been obtained from a statistical survey of proper motions of a large number of the dwarf novae, it is almost certainly more accurate than Hinderer's provisional figure.

Hinderer interpretated his results on the basis of a single star of spectral type dG3 with a surface temperature at minimum of 5,000°K and surrounded by an extensive atmospheric envelope, the radius of which is almost twice that of the star itself. At the time of an outburst, internal explosive conditions in the star bring about a large rise in the surface temperature and an expansion of these outer levels. Once the internal source of energy has been dissipated, the star then returns unchanged to its initial state.

This idea is certainly plausible, although it is one which still does not explain the observed fact that no Doppler shift of the spectral lines occurs as we would expect if there were any appreciable expansion of the outer levels of the star. The alternative explanation, put forward by Joy[2] from his studies on RU Pegasi, that these variables are binary systems is now that which is universally accepted.

Certain differences have also been found by several observers in the spectrum of this star which apparently depend upon the type of maximum. When SS Cygni passed through one of its broad, A-type maxima, the spectral class approximates that of an AO star with the absorption lines visible throughout most of the cycle. At times of a short maximum, however, these absorption features are visible only briefly, appearing only shortly before maximum brightness and fading swiftly as the light begins to decline.

REFERENCES

1 Elvey, C. T., and Babcock, H. W., *Ap. J.*, **97**, 412 (1943)
2 Joy, A. H., *Publ. A.S.P.*, **52**, 324 (1940)
3 Kraft, R. P., *Ap. J.*, **139**, 457 (1964)
4 Greenstein, J. L., *Stars and Stellar Systems*, Vol. 6, p. 676, ed. J. L. Greenstein (Chicago: Univ. of Chicago Press 1960)
5 Copeland, E., *The Observatory*, **5**, 100 (1882)
6 Knott, G., *ibid*, **5**, 110 (1882)
7 Fleming, W. P., *H.A.*, **56**, 210 (1912)
8 Adams, W. S., and Joy, A. H., *Pop. Astron.*, **30**, 103 (1921)
9 Wachmann, A. A., *A.N.*, **255**, 341 (1935)
10 Hinderer, F., *ibid*, **277**, 193 (1949)
11 Kraft, R. P., *Ap. J.*, **142**, 1588 (1965)

4. Dwarf Novae as Binary Systems

In the past two chapters we have considered the light variations and the various changes in the spectra of the dwarf novae which occur throughout the light cycle. From time to time, mention has been made of the binary nature of these variables. Now, we are in a position to combine the information gathered from these two sources in order to put forward a model of a typical dwarf nova. It must be emphasised at the outset that although the picture which will emerge from this discussion explains many of the observations, there are still a large number of features requiring explanation and undoubtedly the presently accepted model will be modified as further data are obtained.

The peculiar light variations of the dwarf novae have, of course, been extensively studied for several decades, yet the physical nature of these variables remained a complete mystery for almost a century after the discovery of U Geminorum. In his excellent monograph on this particular star, written in 1908, Van der Bilt[1] states that 'a physical or chemical explanation of the light changes and the peculiar irregularities of the period must clearly be left to the future'.

In spite of the rapid advances made in both photoelectric and spectroscopic techniques and the commissioning of the 100- and 200-inch reflectors at Mount Wilson and Mount Palomar, a further forty-eight years were to elapse before the long-awaited breakthrough occurred. It is perhaps prophetic that one of the suggestions put forward by Van der Bilt as likely to carry our knowledge of U Geminorum a little further was that, in connection with the possibility that the variable might be a very close double star, special measurements of its coordinates and normal brightness might provide some evidence of this, together with spectroscopic measures of possible variations in radial velocity. It seems likely that he considered there to be a parallel between U Geminorum and the discovery of the companion of Sirius, the sinuous proper motion of Sirius revealing the existence of the companion star long before it was seen visually. We now know that this method would have been utterly useless as far as U Geminorum is concerned. The two components of this variable (and other dwarf novae) are so close that only the spectroscope provides evidence of duplicity.

The actual problem of determining the nature of these peculiar variables is complicated by several factors. Not least of these is the problem of determining the intrinsic luminosity of these stars. To obtain these values we require to know the distances of the dwarf novae and unfortunately the early measurements of the parallax of the brightest member of the class, SS Cygni, were far from being concordant.

Now we have two methods of doing this. One hopeful possibility is to obtain a statistical parallax (although admittedly one subject to some inaccuracy) from the proper motions and radial velocities of a number of these stars. This will, of course, yield a mean figure but at least it will provide us with some idea of the order of distance with which we are dealing and should tell us whether we are concerned with fairly close, faint stars or much more remote, very bright objects. Such a survey has been carried out by Smart[2], the result of this work indicating a mean distance of about 215 light years for four stars; SS Cygni, EY Cygni, U Geminorum and RU Pegasi. From this they would appear to be reasonably near objects with absolute magnitudes at minimum somewhat less than that of the sun.

The second way is to measure the parallax directly for individual variables and over the years this has been done by several astronomers for both SS Cygni and U Geminorum. One of the earlier measurements of SS Cygni gave a value of −0.012 seconds of arc which would suggest that this variable is a quite remote object and therefore highly luminous. Later observations, particularly those made by Strand[3], who found a trigonometrical parallax of +0.032 seconds of arc, lead to an absolute magnitude at minimum of +9.5. A similar estimate of the trigonometrical parallax of U Geminorum leads to an absolute magnitude at minimum of +10.0. Now if we accept these figures and take them in conjunction with the spectra at maximum, we are led to the conclusion that one of the components of these systems is a white dwarf star. Some further confirmation of this comes from an analysis of the colour indices of the dwarf novae at minimum. These range from +0.52 to +0.68 and are very similar to those of certain old novae.

This is all very well in so far as it goes. We are now saying that some features of the dwarf novae are compatible with those found in white dwarf stars and since we shall be using this term quite often now, we must at this point examine the peculiar properties of the white dwarfs in a little more detail.

The Properties of the White Dwarf Stars

The white dwarfs are all intrinsically very faint objects, a fact which explains the relatively small number known and their comparatively recent discovery. On the familiar Russell-Hertzsprung diagram, they occupy a region below and to the left of the main sequence (Fig. 13).

Such stars represent the final stage in the evolution of most stars when all of the hydrogen has become exhausted and even helium burning as a source of energy has virtually ceased. Being so faint, we would naturally expect to find white dwarfs among those stars which are comparatively close to us and except in some special circumstances (such as the old novae and supernovae remnants which are usually quite distant objects), this is indeed the case. The existence of the first known white dwarfs was predicted theoretically some years before they were actually discovered. In 1844, Bessel suggested that the irregularities in the proper motion of Sirius were due to the presence of an invisible companion which has a mass similar to that of the sun and seven years later, Peters calculated the theoretical orbit of this companion from the observed sinuous proper motion of Sirius. It was not until 1862, however, that Alvan Clark found the faint companion visually. Bessel's other prediction concerning a companion of Procyon was later confirmed by its discovery in 1896 by Schaeberle.

At that time there was nothing to indicate that these faint companion stars were significantly different from normal dwarfs, many of which, the red dwarfs, were already known in fairly considerable numbers, most of them being close cosmic neighbours of the sun. However, if we examine these red dwarfs we find that they, like the giant and supergiant stars, all obey the mass-luminosity relation first put forward by Eddington in 1924. One very important characteristic of those stars which obey this relationship (Fig. 14), is that the material of which they are composed behaves as a perfect gas.

The first indication that the companions of Sirius and Procyon are highly abnormal came when their diameters

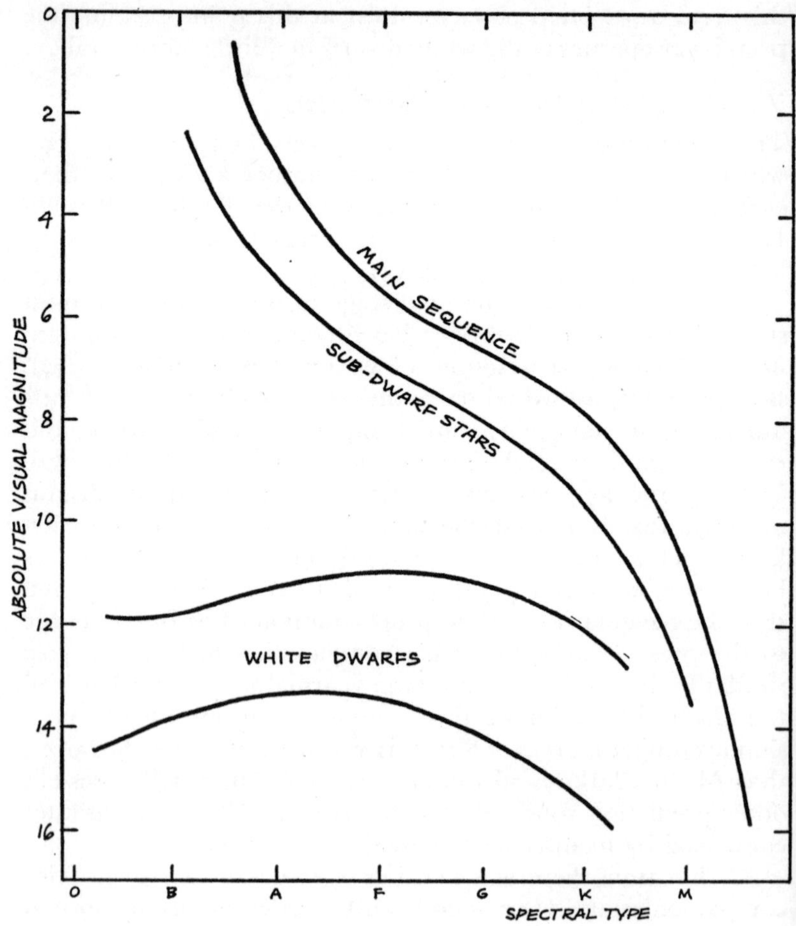

Fig. 13 – The region of the white dwarfs on the Russell-Hertzsprung diagram.

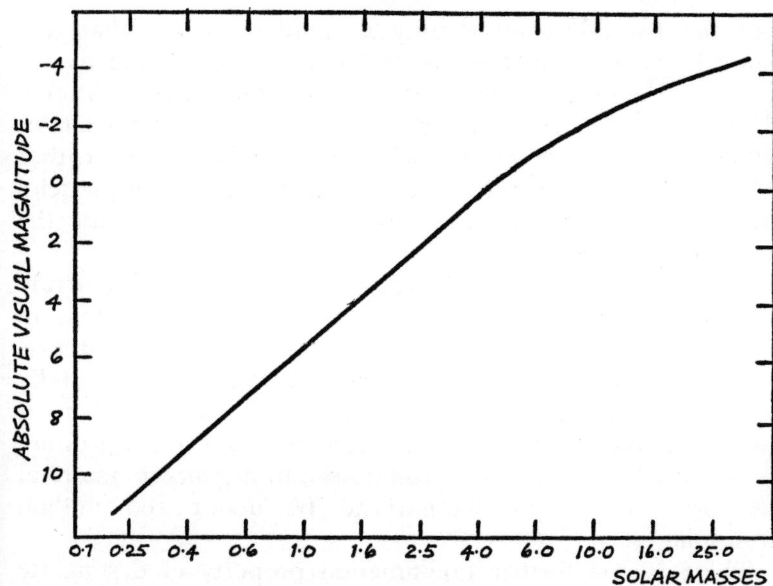

Fig. 14 – The mass-luminosity relationship for main sequence stars.

were determined for this revealed the truly astonishing situation that these stars are fantastically small. In size, they are very like the Earth, yet inside this small volume they contain the mass of a star! As a result, the internal densities of the white dwarfs are unbelievably high, being of the order of 100,000 times that of the densest metal known.

It is scarcely surprising, therefore, that astronomers were extremely loath to accept such a situation. When it later became quite clear that no mistake had been made in the observations, two important questions immediately demanded answers. How can we account for such tremendously high densities? And what is the physical state of the material inside a white dwarf star?

The answers to both of these questions were not long in forthcoming. In 1926, Fowler showed that whereas the matter in the interiors of normal stars continues to behave as a perfect gas in spite of the increase in density towards the centre, this is not the case with the white dwarfs. Under normal conditions of temperature and pressure, even in the centres of stars, the fact that ionisation occurs means that the nuclei of the atoms and the free electrons occupy a smaller volume than the neutral atoms themselves. In the interior of a white dwarf, however, the density is so great that the atomic nuclei and free electrons are so closely packed that the material no longer behaves as a perfect gas. We must bear in mind, however, that even under these incredible conditions, the individual particles do possess certain random motions and so in this sense the material is still a gas.

In its other properties though, material in this degenerate state is very different from normal gaseous matter and in particular, as demonstrated by Chandrasekhar, the greater its mass, the smaller is the volume it occupies. This is a totally paradoxical situation but astonishing as it may be, there is now no doubt at all as to its correctness. The critical pressure above which normal matter collapses into degenerate material has been shown by Kothari to be about 100 million atmospheres.

There is one further fundamental property of degenerate matter which we must consider here. Although the temperature inside white dwarf stars approaches a hundred million degrees, the degenerate matter emits no radiation. If we were able to see it, this 'gas' would appear quite dark. Now this obviously requires some explanation since it appears at first sight to be a complete contradiction. It is easier to visualise this if we rid ourselves of the confusion between temperature and the more general conception of heat. Temperature is nothing more than a measure of the random velocities of the atoms and elementary particles which make up matter. In the case in point, inside a white dwarf star, the electrons are moving with extremely high velocities amounting to tens of thousands of kilometres per second (otherwise the pressures would be quite inadequate to maintain the equilibrium in the star).

Now although the temperature of this degenerate material is so high, the fact that the atomic constituents are packed so tightly together by the tremendous pressure means that the electrons in the atom are all forced into their lowest energy states. Under these extreme conditions there can be no transition from a higher energy level to a lower one and since emission of radiation from an atom only occurs during such electronic transitions, there can be no radiation.

This would seem to be in direct conflict with what we have said earlier, namely that white dwarfs *are* luminous even though they do possess very low intrinsic luminosities. How does this come about if, as we have shown, degenerate matter emits no visible radiation? The answer to this is that a white dwarf does not consist entirely of degenerate matter. There is a very thin outer atmosphere which is non-degenerate for here the tremendously high pressures of the interior are not operative. This outer atmosphere, of course, is only a few hundred kilometres thick and as we shall see in a later chapter, this has a very important bearing on the possible mechanism of the outbursts of the dwarf novae.

Finally, we must examine the evolution of these peculiar stars and in this respect we may obtain quite a lot of information from the Sirius and Procyon systems. Being physical binaries, both components must be of about the same age and from this it follows that the evolution of the white dwarf companion must have been very much more rapid than that of the main component. Clearly, the white dwarf star was initially the more massive of the two for a large proportion of its original mass must have been lost during its evolution. The companion of Sirius, for example, was originally more than 3 times as massive as the sun whereas now it has only 0.9 times the sun's mass. Whatever the evolutionary path may have been, it was obviously one which enabled the star to lose more than two-thirds of its mass. At present, the most logical explanation would seem to be a supernova explosion. There are certain arguments in favour of such an evolutionary career. All of the white dwarfs known closely follow a modified mass-luminosity relationship first enunciated by Chandrasekhar which depends upon the ratio of free electrons to the atomic mass number. For hydrogen this is obviously 1 : 1, whereas for helium it is 1 : 2. When we examine the white dwarfs we find that they approximate much more closely to the latter ratio than the former indicating a great deficiency of hydrogen and a high preponderance of helium. This is exactly the situation we would expect to find following a supernova explosion. There is too, some more direct confirmation for in a few cases we can actually observe the remants of supernovae and we find that these are fairly typical white dwarfs.

On this basis then, we may postulate that at some time in the distant past, the companions of Sirius and Procyon underwent a supernova explosion. Owing to their proximity, this would have been a truly stupendous sight had there been any men on the Earth to see it. Unfortunately, this must have occurred many millions of years before the advent of Man on this planet.

Now what about the white dwarf components of the dwarf novae systems? Are we implying that these stars were produced by a supernova explosion? The answer is no. There is no evidence whatever that the primary components of these binary systems evolved into white dwarfs via the supernova stage. Indeed, most of the available information would appear to be against such an evolutionary scheme. Although both classes of star are clearly of a similar composition, there are certain differences between them. What we may term the normal white dwarfs, represented by the companions of Sirius and Procyon, show no detectable variation in brightness. It may be argued, of course, that stars such as Sirius and Procyon are not only physical, but also optical doubles, that the components are separated by a far greater distance than the components of a dwarf nova system and hence we may expect some interaction between the two stars making up the latter which could account for the light variations. This is certainly a point which must be taken seriously and one which we shall discuss in more detail when we come to consider the nature of the outbursts of the U Geminorum and Z Camelopardalis variables.

Radial Velocities and Orbital Elements of the Dwarf Novae

The spectra obtained by Elvey and Babcock in their survey of these variables were, as we have seen in the last chapter, obtained from long exposures and naturally we would not expect to find any evidence of duplicity or of orbital motion on these spectrograms. Thirteen years later, Joy[4] found the spectrum of SS Cygni at minimum to be composite and was able to show that this variable is a spectroscopic binary with a period of 8 hours and 38 minutes. Here then we have a striking confirmation of Van der Bilt's suggestion that U Geminorum (and similar variables) might be close doubles although as we shall see, the components of these binary systems are much closer together than visualised by him in 1908.

Since Joy's discovery, which was based upon data collected from five spectrograms taken between 1941 and 1955 at times when the star was at minimum, evidence of duplicity has been found in several other dwarf novae, particularly by Kraft[5] and Krzeminski[6]. Indeed, at the present time there is nothing to contradict the theory that all of these variables are binary systems although direct spectroscopic proof of this for the fainter members of the class must await more sensitive techniques. As we saw earlier, in Chapter Two, several of the old novae have been found to be eclipsing binaries, all with very short orbital periods, for example Nova DQ Herculis (1934)[7,8] and Nova T Aurigae (1891)[9].

Of those dwarf novae for which we do have orbital elements, all have periods of less than 9 hours with one single exception. AE Aquarii, which is very atypical of the class, has a period of about 17 hours. Z Camelopardalis, although showing evidence of orbital motion, has not yet had its period accurately determined while EY Cygni has a composite spectrum but any velocity variations are so close to the limit of detection that we

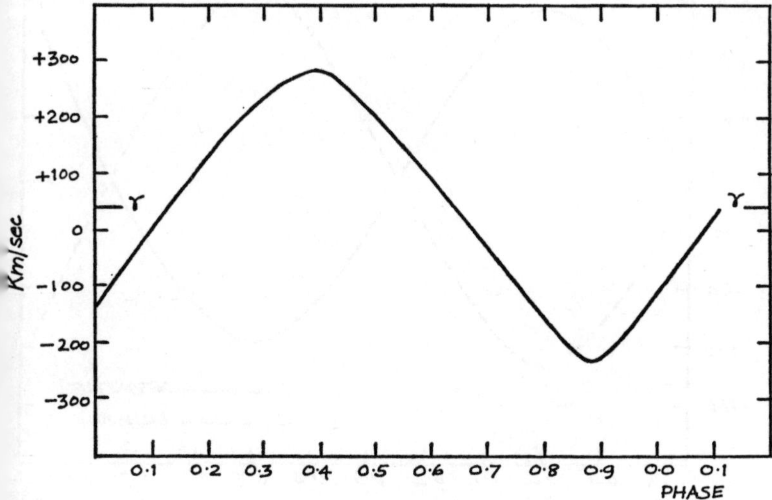

Fig. 15 – Radial velocity curve for U Geminorum obtained from the emission lines (*Astrophysical Journal by kind permission of the University of Chicago Press*).

are still doubtful about their magnitude. It seems very likely that the orbital plane of this variable is so aligned in space that we are viewing it virtually pole-on.

It is a comparatively easy matter to obtain radial velocities from the emission lines of the blue component (Fig. 15), the task of doing this for the late-type star is much more difficult since only in a few cases can the absorption spectrum of this star be differentiated from the broad emission lines due to the bright companion which very often obscure the weaker absorption lines. As we might expect, the most favourable instances for observing this late-type spectrum are where the emission lines are fairly narrow as in the case of RU Pegasi (Fig. 16).

The orbital elements of some of the dwarf novae as determined by Kraft[5] are given in Table 2. As will be seen, the peculiar (γ) velocities are all fairly small, confirming the statistical parallax which led to the value of approximately +9.5 for the absolute magnitude at minimum. Where it has been measured, the mass ratio of the components is very

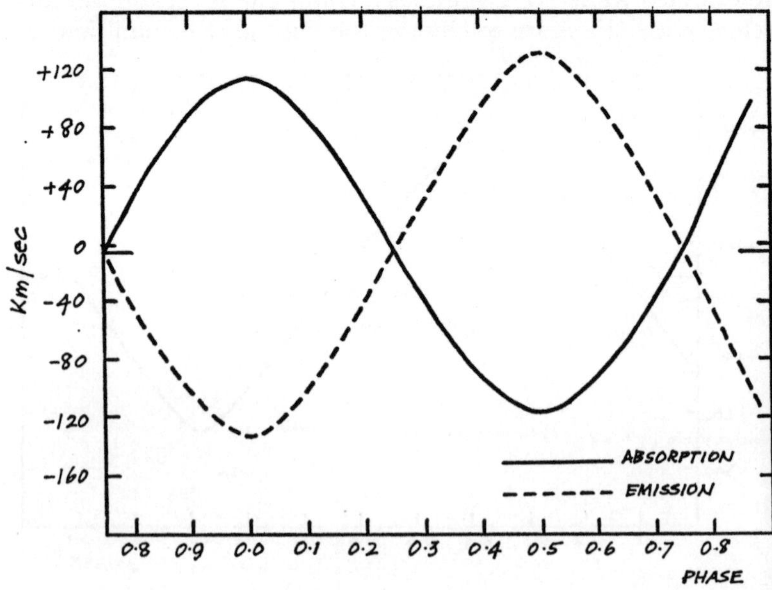

Fig. 16 – Radial velocity curves for RU Pegasi obtained from both emission and absorption lines.

close to unity with the late-type star being slightly the more massive. Although this result has, so far, been obtained for only a few members of the class, it is believed to hold also for the remainder of these variables.

One further point of interest arises from the data given in the table. Whereas the orbits of the U Geminorum variables are essentially circular as determined from their radial velocity curves, that of RX Andromedae (Fig. 61), which is a Z Camelopardalis star, is highly eccentric and it is possible that the same may be true of Z Camelopardalis itself although as mentioned earlier, the radial velocity curve for this particular star has not yet been determined with any degree of accuracy due to difficulties in measuring the velocity variations in the spectral lines. As we shall see in a later chapter, this ellipticity of the orbit of RX Andromedae may provide us with an explanation of the 'standstills' occurring in the stars of this small subgroup.

Table 2

ORBITAL ELEMENTS OF THE DWARF NOVAE

Star	Period h m	K_1^* km/sec	K_2 km/sec	γ km/sec	e	ω	M_1/M_2
RX And	5 05	77.5	—	−18	0.40	220°	—
SS Aur	3 30?	85.0	—	+45	?	?	—
SS Cyg	6 38	122.0	115.0	−09	0.00	—	0.90
EY Cyg	—	?	—	−10	—	—	—
U Gem	4 10	265.0	—	+42	0.05	160°	—
RU Peg	8 54	137.0	112.0	−07	0.00	—	0.85

*The subscript 1 refers to the blue companion, irrespective of whether or not it is the more massive of the two components.

Gas Streams within the Binary System

So far, we have confined ourselves merely to the two components which make up the dwarf novae systems. We shall now examine the third component, namely the luminous stream of gas, chiefly hydrogen, which flows from the cooler star and takes up an orbit around the blue companion. The importance of this gas stream will become more apparent when we come to consider the nature of the outbursts of the dwarf novae for at least one of the theories concerning these supposes that the interaction of this high temperature stream of hydrogen with the degenerate matter of the white dwarf is the basic cause of the outbursts.

Let us first begin by considering the fundamental question – why should a gas stream be present in these binary systems? Is this a situation unique to the dwarf novae or is it a more common occurrence found in other types of binaries? It is now generally believed that such gas streams will be formed in almost every close binary system and although much remains to be understood concerning the dynamics of these rotating streams there are hints as to their formation, the direction of the gas flow and the effect they have on both the orbital period and the synchronisation of axial rotation of the component stars with the period of revolution around the common centre of gravity.

One of the first instances to be discovered of a gas stream within a binary system concerns the variable star RW Tauri. This is a fairly typical Algol variable with a normal photographic magnitude of 8.00, fading by 4.27 magnitudes during the primary eclipse which is total and lasts for 84 minutes. The primary component is a white star of spectral class B9e, the secondary being a larger, yellow star of class K0. As with most of the Algol variables, we are able to observe the spectra of both stars during maximum but the bright emission lines of hydrogen in the spectrum of the B9e star behave in a very puzzling manner. At the beginning of the primary eclipse, these lines are found to be displaced towards the red end of the spectrum indicating that their source is receding from us. At mid-eclipse they vanish altogether only to reappear once more shortly after eclipse but now displaced by an equal amount towards the violet. The explanation of this phenomenon was put forward by Joy who suggested that there is a ring of hydrogen around the primary which is also eclipsed by the secondary component. From the Doppler shift of the emission lines it is possible to calculate the velocity of rotation of this ring. This turns out to be 350 kilometres per second, one revolution taking about 14 hours. The diameter of the ring is approximately 3,540,000 miles.

Similar gas streams have been found, not only in the majority of Algol and β Lyrae variables, but also in the W Ursae Majoris systems which, as has been mentioned earlier, may possibly be the precursors of the dwarf novae. Several of the old novae too, show evidence of such a ring of gas surrounding the hot component and this is probably true also of the recurrent novae but in the case of these latter stars, the task of obtaining this evidence is greatly complicated by the presence of the expanding shell of gas from previous outbursts.

There is now a growing mass of data to suggest that such streams of gas are a normal development in the evolution of close binary systems of every kind. Even in the W Ursae Majoris variables where both components are very similar and are also main sequence stars, it is found that one member of the system is somewhat more massive than the other and consequently the two stars will commence evolving at different times; indeed, it is most likely that one will have completed its evolution to the white dwarf stage before the other even begins to move away from the main sequence!

As far as the dwarf novae themselves are concerned, the presence of a gas stream has been definitely established. One remaining problem, however, is whether the mass which is being ejected into the system from the red secondary, flows continuously through the Lagrangian point, or does so intermittently (possibly only at the time of an outburst). This is a question which is now under active consideration and it has gained far more importance in recent years following the discovery by Krzeminski[10] that the outbursts of U Geminorum are due, not to any change in the blue star, but to loss of mass from the secondary through the inner Lagrangian point. At the moment, it is not possible to give any definite answer to this question. Certainly it would seem that since these late-type stars, which are evolving along a line parallel to the main sequence, have atmospheres in convective equilibrium, there will be irregular oscillations in such stars which will result in an equally irregular ejection of mass.

REFERENCES

1 Van der Bilt, J., *The Variable Star U Geminorum*, Recherches Astronomiques, III (1908)

2 Smart, W. M., *Stellar Dynamics*, p. 208 (Cambridge: Cambridge University Press), (1938)
3 Strand, K. A., *Ap. J.*, **107**, 106 (1948)
4 Joy, A. H., *ibid*, **124**, 137 (1956)
5 Kraft, R. P., *ibid*, **135**, 408 (1962)
6 Krzeminski, W., *ibid*, **140**, 921 (1964)
7 Walker, M. F., *Publ. A.S.P.*, **66**, 230 (1954)
8 Walker, M. F., *Ap. J.*, **123**, 68 (1956)
9 Walker, M. F., *ibid*, **138**, 313 (1963)
10 Krzeminski, W., *ibid*, **142**, 1051 (1965)

5. Mean Absolute Magnitudes

The features of the dwarf novae which have so far been described have been derived by the familiar methods used for the investigation of every other type of variable star. The sample available for the examination of the dwarf novae is very small compared with, say, the large numbers of Algol, β Lyrae, Cepheid and long period variables which are well within the reach of existing instruments. Nevertheless, it provides us with a very welcome, and indeed a necessary, connecting link with other classes of variable with which we will be able to relate the dwarf novae in order to gain further information concerning their physical nature and probable mode of evolution. The light curves of the brighter U Geminorum and Z Camelopardalis variables are fairly comprehensively known thanks to the large amount of observational data which has been accumulated from visual estimates made over the past half century or so; the spectral changes which occur with variations in the phase are, naturally, less fully documented but nonetheless they have furnished the vital information which has led to these stars being recognised as close binary systems.

We must now turn our attention to a further parameter of these variables which has caused a great deal of controversy during the last thirty years or so. Even though it is less well understood than most other features it is, nevertheless, of fundamental importance. This is the absolute magnitude of both the white dwarf primary and the cooler secondary component of these systems. As we shall see, several different methods are available, and have been used, for the determination of absolute magnitudes but unfortunately they have yielded highly discordant results. The absolute magnitudes of the white dwarf components do not present the same difficulty as those of the red secondaries and since, as we shall see later, there is now evidence from the accurate photoelectric work carried out by Krzeminski[1] that it is the secondary which is responsible for the observed outbursts of U Geminorum, it is clearly of great importance that the absolute magnitudes of these late-type stars, and hence their position on the Russell-Hertzsprung diagram, should be known.

Trigonometric Absolute Magnitudes

One quite straight-forward method of obtaining absolute magnitudes, provided the star is not too distant (closer than about 100 parsecs), is by means of the following simple relationship between distance and the apparent and absolute magnitudes.

$$M - m = 5 - 5 \log D \qquad (1)$$

where M is the absolute magnitude, m the apparent magnitude and D is the star's distance in parsecs. Now the apparent magnitude can be readily measured with a high degree of accuracy and if we can, therefore, determine the distance, we have a way of obtaining the absolute magnitude.

The first method to be used for determining the distance to a star was the trigonometric method which is used extensively by surveyors. Since the distances which concern us are far greater than those used in surveying, we must naturally use the longest base-line possible, namely the diameter of the Earth's orbit around the sun, a total distance of 186,000,000 miles (this distance is the astronomical unit and is used in the definition of a parsec which is the distance of a star having a parallax of one second of arc using this base-line). For these stars which lie within 15 parsecs or so of the sun, the trigonometric method has proved to be extremely reliable. Beyond this and up to about 100 parsecs, the accuracy falls off with increasing distance as might be expected since even with the largest instruments and most careful precautions, the difficulties associated with the measurement of the small angles involved become increasingly acute. Once we go beyond 100 parsecs, the method is virtually useless and other means of determining stellar distances must be employed.

As far as the class of dwarf novae is concerned, the nearest members of the group are quite distant objects and trigonometric measurements by different observers have yielded contradictory results. Early determinations of both SS Cygni and U Geminorum were made by van Maanan[2]. The value obtained for SS Cygni was $-0''.012 \pm 0''.012$ which, as we saw in the previous chapter, would suggest that this dwarf nova is a very distant and highly luminous object. For U Geminorum, on the other hand, van Maanan found a trigonometric parallax of $+0''.010 \pm 0''.012$, the size of the probable error being greater than the measured parallax itself. Although this would indicate that little reliance can be placed upon the latter determination, the calculated absolute visual magnitude at minimum of $+9.1$ is in good agreement with the value of $+9.5$ which we obtain from the figure of $+0''.032 \pm 0''.010$ found by Strand[3] for SS Cygni. Now from what we know of the secondary components of these dwarf nova systems, there would appear to be no reason to suppose that they form a very heterogeneous group and accordingly we would expect to find their absolute visual magnitudes at minimum to be very similar. Let us now take a look at a second method for determining absolute magnitudes.

Spectroscopic Absolute Magnitudes

The evidence presented in the previous chapter establishes fairly well that the dwarf novae are binary systems having orbital periods between three and about fifteen hours. For most of those which have been examined spectroscopically, the spectrum indicates the presence of a hot subdwarf star, very similar to the white dwarfs and clearly in an advanced state of helium burning. In a few of them (those which have orbital periods in excess of about 6 hours), the absorption spectrum of a late-type dwarf is also visible, for example AE Aquarii, SS Cygni, EY Cygni, RU Pegasi and Z Camelopardalis. In those variables where the late-type spectrum is not visible and the orbital period is shorter than 6 hours, we may presume that the secondary component is relatively fainter compared with the primary and, as seems most likely, of an even later spectral class than dG5 to dK0 found from those systems in which both spectra are visible.

From the spectral classification of those variables in which it is possible to determine the secondary spectrum, Kraft has found an average absolute magnitude for the secondary components of +5.4 which, when corrected for that of the primary star gives a mean figure of +4.8 for the systems themselves. This wide gulf between the trigonometric and spectroscopic absolute magnitudes attains serious proportions when we come to plot the positions of both primary and secondary components on the Russell-Hertzsprung diagram. If we accept the fainter absolute magnitudes given by the former method, we find both components lying well below the main sequence. Alternatively, the spectroscopic absolute magnitudes place the secondary components either on the main sequence or just to the right of it.

Several attempts have been made to resolve this grave difficulty. The trigonometric values are clearly open to criticism on the grounds that not only are the results obtained by different observers conflicting (almost certainly due to the problem of measuring these very small parallaxes) but the number of determinations is extremely small, quite insufficient for any great weight to be placed upon them. Is the position any better as regards the spectroscopic values? Even here, more recent work has indicated that there is a good deal of uncertainty surrounding these figures. We have already seen in Chapter Three that the luminosity class of RU Pegasi is not constant, but varies throughout the orbital period from Class V to III, being highest at the phase when the inner Lagrangian point is nearest to the observer. This would tend to suggest that perhaps the luminosity class which we have assigned to the secondary components is too high. If this is so, then the true luminosity of these stars, being lower than that determined directly from the spectrum, will yield absolute magnitudes somewhat fainter than +5.4 and the gap between the spectroscopic and trigonometric absolute magnitudes is reduced.

Why should there be this difference between the actual and observed luminosities of these secondary components of the dwarf nova systems? This is not an easy question to answer. We have, of course, been assuming that this component behaves in the same manner as a single star of the same spectral class. Are we, however, justified in making this assumption? We must not lose sight of the extreme complexity of these binary systems and especially the fact that the secondary fills its lobe of the equipotential surface. As Kraft[4] has pointed out, since this is the case, the gravity of such a star will be somewhat less than that of a non-rotating star having the same mass. This will certainly result in the apparent luminosity being higher than is actually the case. We must also take into account the possibility that the presence of the hot white dwarf companion may have some effect on the radiation field of the secondary as suggested by Kraft and Luyten[5] although at the moment, it is difficult to estimate what the full effect of this will be. From the foregoing discussion, however, it is clear that, as in the case of the trigonometric results, it would be unwise to place too much reliance upon the spectroscopic absolute magnitudes which have so far been obtained.

Absolute Magnitudes Based upon Statistical Parallax

Several attempts have been made to derive a statistical parallax for the dwarf novae. One of the first was made by Smart[6] who obtained a mean distance of 215 light years for the U Geminorum variables, his survey including four stars; SS Cygni, EY Cygni, U Geminorum and RU Pegasi. The average absolute magnitude at minimum found from this work is +6.0, slightly fainter than that of the sun.

Now for an accurate statistical parallax, we need to know two quantities, namely the proper motions and the radial velocities. As long ago as 1934, Kukarkin and Parenago[7] measured the proper motions of both SS Cygni and U Geminorum and showed that such measurements were quite feasible. Further similar observations were carried out by Miczaika and Becker[8] and Mannino and Rosino[9]. When we come to examine the radial velocities, however, the position is not quite as straight-forward. Being binary systems, their radial velocities will vary over a period of a few hours and in addition we have to face the problem of their extreme faintness when at minimum.

In spite of these difficulties, the radial velocities of eleven dwarf novae have been measured with a fair degree of accuracy (Table 3).

The proper motions of twenty-five dwarf novae, including a few of the southern members of this class, have been measured by Kraft and Luyten[5] together with a direct measurement of the solar motion from the radial velocities of these stars. The

Table 3
RADIAL VELOCITIES OF DWARF NOVAE

Star	Radial velocity (km/sec)
HV 8002	+30
RX And	−18
SS Aur	+33
U Gem	+42
Z Cam	−33
T Leo	+11
EY Cyg	−10
WZ Sge	−25
AE Aqr	−53
SS Cyg	−09
RU Peg	−07

solar motion thus derived indicates a direction which is in good agreement with that found by Vyssotsky[10] while the high value obtained for the solar motion (admittedly somewhat uncertain) of +46.1±14.0 kilometres per second, is consistent with the idea that the motions of the dwarf novae are those of typical G- and K-type dwarf stars. The average absolute magnitude at minimum for the twenty-five dwarf novae examined is found by this statistical method to be about +7.5.

The very marked discrepancy between this result and that obtained earlier from the spectral classes of the secondary components is at once obvious. Since the figure of +7.5 magnitude has been obtained using a much larger sample, we would be justified in accepting it in preference to the much brighter absolute magnitude of +4.8. Kraft and Luyten[5] have suggested that this apparent anomaly may be removed if we assume that, although all of the red components of the dwarf nova systems lie on the main sequence, their temperatures and radii remain in constant equilibrium with their masses. As a consequence of this, we find that the absolute magnitudes of these variables at minimum brightness vary over a range which is probably as large as 5 magnitudes, the absolute magnitude of the secondary decreasing more rapidly than that of the blue primary as the orbital period becomes shorter from one variable to the next.

The Period-Luminosity Relationship

From what has just been said, there is clearly support for a period-luminosity relationship among the members of the dwarf novae. As we shall be discussing this particular point fully in Chapter Seven, we will defer it until then.

REFERENCES

1 Krzeminski, W., *Ap. J.*, **142**, 1051 (1965)
2 Van Maanan, A., *ibid*, **87**, 424 (1938)
3 Strand, K. A., *ibid*, **107**, 106 (1948)
4 Kraft, R. P., *ibid*, **135**, 408 (1962)
5 Kraft, R. P., and Luyten, W. J., *ibid*, **142**, 1041 (1965)
6 Smart, W. M., *Stellar Dynamics*, **208** (Cambridge, Cambridge University Press), (1938)
7 Kukarkin, B. V., and Parenago, P. P., *Peremennye Zvezdy*, **4**, No. 8 (1934)
8 Miczaika, G., and Becker, W., *Heidelburg Veröff.*, **15**, No. 8 (1948)
9 Mannino, G., and Rosino, L., *Padua Publ.*, No. 14 (1950)
10 Vyssotsky, A. N., *Publ. A.S.P.*, **69**, 109 (1957)

6. Masses of the Dwarf Novae

Since, as now seems highly probable, the dwarf novae are all close binary systems, we are evidently in a position to make some fairly accurate determinations of the total masses of these systems and hence those of the individual components. If these variables were single stars like the Cepheids, for example, we would of course, have no direct means of calculating their masses and would be forced to rely entirely upon indirect methods such as the position of the stars on the familiar Russell-Hertzsprung diagram where it is often possible to utilise the well-known mass-luminosity relationship provided that the star does not lie too far from the main sequence. Since there is some observational evidence that the primary components, at least, lie well below the main sequence, this method would not be really satisfactory.

In the case of the eclipsing binaries, we have an excellent way of deriving their masses quite accurately since the motion of these stars around their common centre of gravity conforms with the universal law of gravitation and we are able to deduce their masses by means of Kepler's third law in much the same way as that of a planet may be determined from the motion of its satellites. Many of the Algol and β Lyrae variables have had the masses of their components calculated in this way.

Can we apply similar methods for the determination of the masses of the two stars which make up a dwarf nova system? At first sight, it may seem there should be little difficulty in doing so, but unfortunately when we come to apply this method to the dwarf novae, we find the position is appreciably more complex. Not only is there the complicating presence of the circulating gas stream to be taken into account but very few of these variables have an orbital inclination close to 90°; that is, with the plane of the orbit aligned with the line of sight as seen from Earth. Even in those favourable cases where an eclipse has been demonstrated, there is no indication of a secondary eclipse and the primary minimum has an amplitude of only a few tenths of a magnitude. By comparison, the minima in the light curves of the Algol and β Lyrae eclipsing variables have amplitudes often amounting to two or more magnitudes.

Since the spectra of both components are visible in the photographic region of the spectrum in the case of SS Cygni and RU Pegasi, it is possible to obtain radial velocity curves for both the primary and secondary components from the emission and absorption lines respectively and from these, and the corresponding orbital elements, we may obtain an estimate of their masses. Owing to uncertainties in the measurements, there are similar uncertainties in the calculated masses. Kraft[1] has estimated these to be 0.18 and 0.27 solar masses for the blue primaries and 0.20 and 0.32 solar masses for the late-type companions respectively. As we shall see later, these estimates are almost certainly too low.

Orbital Inclinations

Before going on to discuss another, more indirect, method of determining the masses of the components of the dwarf nova systems, it will be helpful first of all to examine the question of the orbital inclination more thoroughly since this will afford us a much clearer insight into the problem of observing eclipses in these systems as well as providing us with an indication of their respective masses. The estimation of the orbital inclination is a comparatively simple matter in the case of the Algol and β Lyrae variables. Since all of these stars *are* eclipsing variables, this immediately tells us that the orbit must be inclined very nearly, if not actually, with the line of sight as seen from Earth. In addition, the amplitude of the primary eclipse is normally large and in most cases, the secondary eclipse can be fairly readily measured although, naturally, its amplitude is somewhat smaller. The shape of the minima too, tell us whether the eclipse is total, annular or partial and since the dimensions of the components may be determined from the elements of the orbit, we may derive the inclination directly.

There is, of course, one basic difference between the eclipsing variables we have just mentioned and the dwarf novae. Fundamentally, the latter variables are not regarded as eclipsing stars, belonging to the large class of eruptive variables. Indeed, until a very few years ago, they were not even considered as binary systems at all. The main characteristics of their light curves are, as we have seen, typically those of a nova-like star and totally unlike those of the Algol variables. The fact that a very small number are eclipsing systems is one which cannot be determined from their visual or photographic light curves and only comes to light when sensitive photoelectric determinations are made during the time one or two of them are at minimum brightness.

Consequently, it is a far from simple matter to obtain accurate figures for the orbital inclination of the dwarf novae. Certainly there is no doubt that U Geminorum has an orbital inclination of approximately 90° and the same appears to be true of Z Camelopardalis for here we have observational evidence of an eclipse in these systems although, as we saw in the previous chapter, the amplitude is very small and no secondary eclipses have been detected, even with the aid of the most sensitive equipment and the highest gains compatible with the signal-to-noise ratio. Kraft has also shown[1] that the primary eclipse varies both in depth and shape, a result which Krzeminski[2] has found to be due to the fact that during the time Kraft made his observations, the star in question, U Geminorum, was increasing in overall brightness at the beginning of a rise to maximum. The U Geminorum variable, EX Hydrae, also may have an orbital inclination very close to 90° but in none of the three systems mentioned, has it proved possible to photograph the absorption spectrum of the secondary star.

The fact that we can photograph the absorption spectrum of the late-type component in the case of both SS Cygni and RU Pegasi indicates that these two variables are the best candidates for an investigation of this kind and both have been examined by Kraft[1]. As a first step, let us take a look at the radius of the cool secondary in each case and compare this with the corresponding lobe of the equipotential surface (we have already seen that several lines of argument lead us to believe that the blue component is a fairly typical white dwarf star, possibly a little closer to the main sequence than the Companion of Sirius, but still quite small and certainly not large enough to fill its lobe of the equipotential surface).

From the energy distribution in the continuum of the spectrum of the variable when at minimum, we can estimate the spectral types of the secondary components of these two systems, these being dG5 and dG8 respectively. Now although these stars have undoubtedly begun to evolve away from their initial positions on the main sequence, as shown by the fact that, as we shall now prove, they have completely filled their limiting surfaces and are ejecting mass through the inner Lagrangian point, we know from other considerations, especially from our knowledge of other stars of these spectral types, that the average radii of dG5 and dG8 stars are 6.2×10^{10} and 5.9×10^{10} centimetres.

The next problem is to calculate the effective radii of the larger lobe of the equipotential surfaces for the SS Cygni and RU Pegasi systems. This we may do by means of the following equation which is due to Kuiper and Johnson[3].

$$\log \frac{r_2}{d} = -0.335 \log \frac{\mathfrak{M}_1}{\mathfrak{M}_1 + \mathfrak{M}_2} - 0.511 \qquad (1)$$

where r_2 is the radius of the larger inner lobe, d is the distance between the centers of mass, $\mathfrak{M}_1$ is the mass of the blue primary and $\mathfrak{M}_2$ the mass of the secondary component. In neither of these systems do we know the value of d accurately but it is possible to assign minimum values for it, these being 9.00×10^{10} and 12.75×10^{10} centimetres for SS Cygni and RU Pegasi. Hence, as Kraft[1] has shown, the minimum values of r_1 for SS Cygni and RU Pegasi are 3.6×10^{10} and 5.1×10^{10} centimetres respectively. Since these figures are less than the estimated radii of the stars themselves we may assume that in both instances, our original statement that the late-type components completely fill their lobe of the equipotential surface is correct. If, now, we equate the two radii we have thus obtained, that of the star itself and that of the larger inner lobe, we may obtain a rough idea of the inclination of the orbit. This works out at about 35° and 60° for these two variables.

At this stage, it is appropriate to consider whether we have any means of checking these figures. Clearly, taking into account the radius of the late-type companion, if the orbital inclination is greater than a certain critical value there ought to be some evidence of eclipse, partial at first and then approaching total as the inclination approaches 90°. For stars of the relative sizes we find in the dwarf novae, this critical value below which there will be no eclipse of the primary, is about 75°. U Geminorum, as we have already seen, in company with Z Camelopardalis and EX Hydrae, have orbital inclinations very close to 90° and eclipses have been demonstrated in these systems.

From accurate photoelectric studies of SS Cygni, Grant[4] has shown that no eclipses occur. In the other case of RU Pegasi, we do not yet know whether or not this is an eclipsing system. If our rough estimate of about 60° for the inclination of the orbit is close to the true figure, then since this lies below the critical value of 75°, we would not expect to find any indication of eclipse. Future work will doubtless settle this issue although there is the added complication of the very close optical companion of approximately the same magnitude as RU Pegasi when at minimum.

Orbital Inclinations Derived from a Study of the Rotating Gas Stream

We now come to a second method by which we can tell whether the inclination of the orbit deviates appreciably from 90°. This depends upon the fact that there is a rotating gas ring in the systems. There is now considerable observational evidence that the secondary stars of the dwarf novae lie on the main sequence and have entered an evolutionary phase which results in them moving down the main sequence rather than to the right of it. In any event, these stars have begun hydrogen burning in their centres, resulting in an increase in the helium core and subsequent expansion of their overall dimensions. Consequently there is a continuous ejection of mass (in the form of a stream of hydrogen-rich gas) through the inner Lagrangian point. This gas takes up the shape of a disc around the blue companion (almost certainly it does not result in the formation of a spherical shell). Now this material itself possesses quite a high temperature and, as was mentioned earlier in the case of the Algol variable RW Tauri, this ring will give rise to a series of bright emission lines in the spectrum, lines which may be comparatively easily differentiated from those produced by the white dwarf star. The spectral lines will appear in emission since they originate in a very hot gas and do not pass through any absorbing medium before reaching us.

The importance of this observation is that, irrespective of the optical thickness of this gas, if the inclination of the ring is close to 90° then these emission lines should be doubled since one edge of the ring will be moving away from us while the other edge will be approaching, resulting in the familiar Doppler shift of the lines in the spectrum. This characteristic doubling of the emission lines associated with the gaseous ring has been observed in the case of U Geminorum where it is very conspicuous and, as we have indicated earlier, there are also suggestions of it in the spectra of Z Camelopardalis and EX Hydrae. By the same argument, of course, if the orbital inclination departs appreciably from 90° there will be no detectable Doppler shift (since this is only found when the motion of the source is directly towards or away from us) and the emission lines due to the ring will thus appear single; this indeed being so for both SS Cygni and RU Pegasi.

It is worth mentioning at this point that observations carried out spectroscopically by Greenstein and Kraft[5] on the old nova DQ Herculis (1934) have shown the presence of certain emission lines (those of the higher Balmer series of hydrogen) which originate from a region in the system in the vicinity of the blue star and are also definitely double in character. These lines also show a superimposed radial velocity which is similar to that of the blue star as it moves in its orbit and they have been interpreted on the basis of a ring of gas surrounding the white dwarf component. DQ Herculis, however, presents a more difficult problem when it comes to an interpretation of the system than, say, U Geminorum, in that this old nova also contains a second nebulous region which produces its own set of emission lines; this being an expanding shell of gas which is almost certainly the remnant of the 1934 outburst.

Using the data obtained from this study of the orbital inclinations, Kraft[1] has estimated the masses of the late-type companions of the SS Cygni and RU Pegasi systems as being about 1.5 times that of the sun, a figure which is certainly of the right order although as we shall see later in the chapter, it may be a little on the high side. We must now take up the important question of the masses of the blue primary components in these binary systems.

Masses of the White Dwarf Components

In 1941, Kuiper[6] showed that an atom in the gas stream present in these close binary systems approximately conserves its angular momentum with respect to the white dwarf star as it moves from the inner Lagrangian point to take up its orbit around this star. Using this information, Kraft[1,7] has obtained a relationship relating the masses of the two components, the orbital period and the width of the emission lines in the spectrum.

$$\frac{G\mathfrak{M}_1 \sin^3 i}{\nu \sin i} = \lambda\ (d_1 \sin i)^2\ \left(1.39\ \frac{\mathfrak{M}_1}{\mathfrak{M}_2} - 0.39\right) \frac{2\pi}{P} \qquad (2)$$

In the above question, $\nu \sin i$ is the projected half-half-width of the emission lines, λ is a proportionality factor which is included to allow for the fractional change of angular momentum and $\mathfrak{M}_1$ and $\mathfrak{M}_2$ are the masses of the primary and secondary components respectively. From a study of SS Cygni and RU Pegasi, Kraft[1] has found that λ has the values 1.26 and 1.36 respectively and we may therefore take a mean figure of 1.31 and apply this to the other dwarf novae. In particular, for U Geminorum itself, we may write.

$$\frac{G\mathfrak{M}_1}{\nu} = 1.31\ d_1^2\ \left(1.39\ \frac{\mathfrak{M}_1}{\mathfrak{M}_2} - 0.39\right) \frac{2\pi}{P} \qquad (3)$$

The second term on the right hand side of the equation is simplified since in this case sin i may be taken as unity (the inclination of the orbit is approximately 90°). For this particular variable, d_1 is found to be 6.31×10^{11} centimetres and the orbital period P is 1.50×10^4 seconds. We do, however, run into a little trouble when we come to decide the value to be assigned to ν, for we find that this parameter is strongly dependent upon whether the emitting region (that is the disc of gas formed around the blue star) is optically thin or thick. Clearly this will have a very profound effect upon the masses calculated by means of equation (3).

Kraft[1] has pointed out that if we consider this region to be optically thin, then the value of v which is obtained, 670 kilometres per second, gives a mass for the blue star of between 1.9 and 5.5 solar masses, the mass ratio of the two stars lying between 0.8 and 1.9. Does this figure agree with observation? Unfortunately, it is far too high since, as Schwartzchild[8] has demonstrated, this star is certainly a white dwarf and even though we may argue that it differs in certain respects from typical members of this group, its mass cannot be much greater than the theoretical upper limit for these stars of 1.2 solar masses. Perhaps, then, we have made the wrong assumption and we should take the alternative case and regard the emitting region as being optically thick. If we do this, the value of v is less than 670 kilometres per second. Unfortunately, in this case, the mass of the blue component turns out to be even higher!

Quite obviously, we must regard the emitting region as being optically thin and taking all considerations into account, it is generally agreed that the masses of the blue stars in the dwarf nova systems lie between 0.7 and 1.1 solar masses. Such a result agrees with the theoretical arguments regarding the masses of white dwarf stars and if we assume a similar range of masses for the secondary components, the result ties in very well with the figures obtained previously for the inclinations of the orbits. When we come to discuss the possible evolution of the dwarf novae in a later chapter, it will become apparent that these considerations have an extremely important bearing on the various theories which have been advanced.

To complete the discussion as far as the whole class of the dwarf novae is concerned, we must, of course, make some mention of the Z Camelopardalis variables. As we have seen, they form only a very small subgroup and it is indeed fortunate that one of them, Z Camelopardalis itself, also shows the spectra of both primary and secondary stars as found by Kraft, Krzeminski and Mumford[9]. We are therefore in a position to make an estimate of the minimum masses and mass ratio of the components from the spectrum of this star. The calculated values of 0.565 solar masses for the white dwarf component and 0.660 solar masses for the secondary star are very similar to those found for the U Geminorum variables just discussed.

Our discussion in Chapter Two described the very close

similarities between the light curves of the dwarf novae, normal novae and the recurrent novae and in the next chapter we shall be examining in detail the evidence which has led some astronomers to postulate an even closer generic relationship between these three classes of variable. It is therefore important that we should now look at the masses of the components of the novae and recurrent novae and see how they compare with those of the U Geminorum and Z Camelopardalis stars just discussed. We have already seen in Chapter Three that the spectra of the old novae differ in certain important details from those of the dwarf novae, chiefly in their wider diversity and the higher excitation of the emission lines, the latter indicating that, in general, the blue components of these systems are hotter than the corresponding stars of the U Geminorum variables. In addition, the physical nature of the dwarf nova outbursts is clearly not the same as that in the novae and recurrent novae so that although all three classes appear to obey the same cycle-amplitude relationship, there are obviously some differences to be expected, differences which may show up in their masses.

Masses of the Novae

Several of the old novae have been examined by Kraft[10] using the 200-inch Palomar reflector with spectroscopic facilities capable of reaching photographic magnitude 15.5 with an exposure time of only 40 minutes. Such a short exposure is necessary with many of these binary systems since the orbital period is so short and longer exposures would result in an inadequate time resolution. Of the six old novae investigated; GK Persei (1901), T Aurigae (1891), DQ Herculis (1934), V603 Aquilae (1918), V841 Ophiuchi (1848) and DI Lacertae (1910), it was possible to obtain estimates of the individual masses of the components for only two of these – GK Persei and DQ Herculis. The blue and red components of GK Persei have masses about 1.29 and 0.56 times that of the sun, those of DQ Herculis being somewhat smaller – approximately 0.12 and 0.20 solar masses. The red component of GK Persei is also noteworthy in that the luminosity class varies erratically from Class V to Class III, being somewhat similar to that of the U Geminorum variable RU Pegasi in this respect.

Mumford[11] has demonstrated that this star is not an eclipsing variable like DQ Herculis and it is quite possible that the masses given above may be a little lower than the actual masses. GK Persei is also of interest since its orbit is highly eccentric, being the only other eruptive variable to show this feature apart from RX Andromedae. The spectral class of the secondary is also abnormal. Certain features would appear to indicate a type between K0 and K5, while others are more suggestive of spectral class G8.

Although the amount of data available is very small, it would appear that the masses of the components in old nova systems cover a much wider range than those present in the dwarf novae. We shall now see how this range (and also that of the orbital periods) is enormously increased when we come to include the recurrent novae.

Masses of the Recurrent Novae

Four of the recurrent novae have been studied in detail by Kraft[10] under the same conditions as described above for the old novae; V1017 Sagittarii (1901, 1919), RS Ophiuchi (1898, 1933, 1958, 1967), T Coronae Borealis (1866, 1946) and WZ Sagittae (1913, 1946). Of these, estimates of the individual masses of the components have been made for only the last two. V1017 Sagittarii shows a very strongly developed absorption spectrum upon which is superimposed extremely weak emission lines. Unfortunately the variable lies to the south of the galactic equator and is also very faint. RS Ophiuchi, however, is relatively bright at minimum (photographic magnitude 11.5) but here the spectrum is exceedingly complex and it would seem that the absorption lines in the spectrum correspond, not to the secondary star but to the presence of an absorbing shell in the system. There is also the added difficulty of the erratic behaviour of the emission lines which make it virtually impossible to elucidate any orbital motion at all.

T Coronae Borealis was examined spectroscopically by Sanford[12] in 1949 using long exposures at the coude focus of the 100-inch reflector. The absorption lines are those of a red giant star of spectral type gM3 which is itself irregularly variable as might be expected from this type of star. Sanford was able to show that the secondary component has a velocity variation of about 230 days, possibly the longest orbital period of all the cataclysmic variables. At the time of his investigation, however, the expanding shell of gas from the 1946 outburst still produced its own emission lines and it was therefore impossible to detect any orbital motion of the primary component from the emission lines.

Kraft[13], however, was able to do this some ten years later when the gaseous shell had dissipated somewhat and found minimum masses of 2.1 and 2.9 solar masses for the primary and secondary components respectively. If we make the assumption that, like the secondaries in the dwarf novae, the red star in this system also fills its lobe of the equipotential surface, then the orbital inclination turns out to be close to 70° and the deduced masses of the blue and red stars become 2.6 and 3.7 solar masses.

Evidently the T Coronae Borealis system is far different from the others we have just been considering. The mass of the blue star is clearly far greater than the theoretical upper limit for white dwarfs and yet, as Kraft has shown, there are no errors in the velocity curve which will alter this conclusion. We are therefore faced with the conclusion that this star is not a white dwarf in the strict sense of the term. The secondary too, is most unusual for these systems. Judging from its size, it must have evolved to the right of the main sequence, now lying in the region occupied by the irregular variables, almost all of which are red giants or supergiants. This is borne out by the irregular light variations of this recurrent nova at minimum which have been shown to be due to the secondary and not to the hot primary which is the seat of the much greater nova outbursts. The orbital period, too, is very long and all in all, this is an extremely anomalous system as far as the cataclysmic variables go.

When we come to consider WZ Sagittae, we find certain anomalies here also. The absorption lines are those of a typical white dwarf as pointed out by Greenstein[14] while the hydrogen emission lines are distinctly double. As we have seen earlier, this suggests that the star may be an eclipsing binary and Krzeminski[15] has shown this to be so, the orbital period being one of 81.5 minutes. The light curve of the star at minimum brightness is very like that of a typical W Ursae Majoris variable, a type of star which we shall be discussing in detail later in the book in connection with the theory that the dwarf novae are a later stage in the evolution of these stars. Kraft[16], using the technique of trailing continuously over a long slit, has found that an S-wave is superimposed upon the doubled hydrogen emission lines, this wave having a period of 80 minutes, almost identical to the orbital period. There now seems little doubt that the mass of the primary in this system, like that of T Coronae Borealis, is too high for it to be a white dwarf.

The smallest mass, calculated on the assumption that the S-wave is associated with a rotating ring of gas with a relatively dense portion in it, yields a value of 1.62 solar masses. On the other hand, the masses of both primary and secondary may be as high as 6.5 solar masses.

There are other differences of detail as far as this particular star is concerned, differences which suggest that, in spite of the long period between major outbursts, it is more like a dwarf nova than a recurrent nova. The absolute magnitude at minimum would seem to be very faint. Greenstein[14] has suggested a figure of +10 and this, together with the absence of the emission lines of singly ionised helium in the spectrum at minimum, is indicative of the dwarf novae rather than the recurrent novae.

It is now appropriate to give a summing up of the discussion in the present chapter. It will be clear that as far as the dwarf novae are concerned, the range of masses for both primary and secondary components is quite narrow, between 0.7 and 1.1 solar masses. By accepting the errors which may be present in our calculations, we may also assume that the blue primaries all lie below the theoretical upper limit set on the masses of white dwarf stars. When we consider the novae, and in particular the recurrent novae, the position is one of much greater heterogeneity and we are faced with the problem of much larger masses for the hot stars in these systems than are compatible with their classification as white dwarfs.

REFERENCES

1 Kraft, R. P., *Ap. J.*, **135**, 408 (1962)
2 Krzeminski, W., *ibid*, **142**, 1051 (1965)

3 Kuiper, G. P., and Johnson, J. R., *ibid*, **123**, 90 (1956)
4 Grant, G., *ibid*, **122**, 566 (1955)
5 Greenstein, J. L., and Kraft, R. P., *ibid*, **130**, 99 (1959)
6 Kuiper, G. P., *ibid*, **93**, 133 (1941)
7 Kraft, R. P., *ibid*, **139**, 457 (1964)
8 Schwartzchild, M., *Structure and Evolution of the Stars* p. 233 (Princeton: Princeton University Press) (1958)
9 Kraft, R. P., Krzeminski, W., and Mumford, G. S., *Ap. J.*, **139**, 678 (1964)
10 Kraft, R. P., *ibid*, **130**, 110 (1959)
11 Mumford, G. S., *A.J.*, **68**, (1964)
12 Sanford, R. F., *Ap. J.*, **109**, 81 (1949)
13 Kraft, R. P., *ibid*, **127**, 625 (1958)
14 Greenstein, J. L., *ibid*, **126**, 23 (1957)
15 Krzeminski, W., *Publ. A.S.P.*, **74**, 66 (1962)
16 Kraft, R. P., *Science*, **134**, 1433 (1961)

7. Relationship to the Novae

Reviewing our picture of the dwarf novae, in particular the brighter members whose light curves have been most thoroughly observed, it is scarcely surprising that a generic relationship was sought between the U Geminorum variables and the normal novae. Indeed, the discoverers of most of the early members of this class (and here we also include the Z Camelopardalis variables) considered them as possible novae and it was not until later observations provided proof of their short periods that they were recognised as a distinct class of variable star. In the present chapter we shall be discussing the many points of similarity between the two groups, together with those areas where they show obvious dissimilarities.

The general light variations of the dwarf novae are now reasonably well known, characterised by a usually abrupt rise of between two and six magnitudes, a much slower decline and fairly smooth minima (for the purpose of this discussion we shall ignore the irregularities associated with the Z Camelopardalis variables). Such light changes are also characteristic of the novae. The latter do, of course, have larger amplitudes and far longer periods running into hundreds or thousands of years. Strictly speaking, the small class of recurrent novae is more closely allied to the U Geminorum stars in this respect and as we have already seen the cycle-amplitude relation for these stars is the same as that for the dwarf novae. Quite clearly, as far as the superficial light variations are concerned, the maxima are extremely alike, differing only in that there would appear to be a larger variety among the light maxima of the novae – they do not form so homogeneous a group as the U Geminorum stars as may be illustrated by the light curves of the two novae given in Fig. 17. Nova Delphini (1967), discovered by Alcock, has proved to be one of the slowest on record, certainly the most remarkable of the present century. The presence of secondary outbursts is well shown in the light curve.

Spectroscopic Similarities

The physical and spectroscopic characteristics of several old novae when at minimum light have been comprehensively investigated by many astronomers, notably by Kraft[1,2], Walker[3,4,5,6,7], Greenstein[8,9] and Wallerstein[10]. The recurrent nova T Coronae Borealis (1866 and 1946) has also been studied in detail by Kraft[1] and Sanford[11].

Following the discovery in 1954 by Walker[3] that Nova DQ Herculis (1934) is a binary system there was naturally considerable speculation as to whether all novae might be binaries

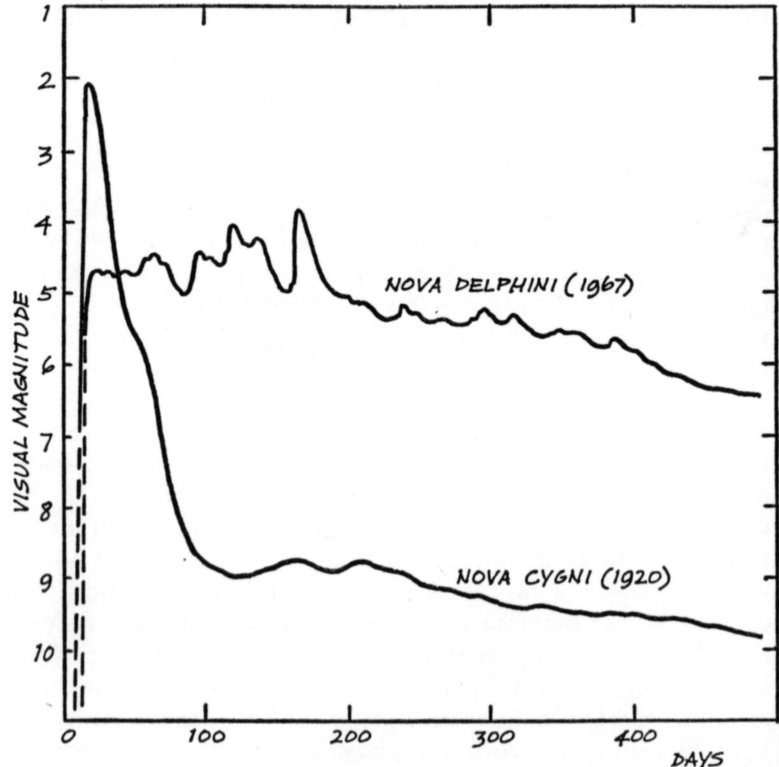

Fig. 17 – Light curves of Nova Cygni (1920) and Nova Delphini (1967) (*Courtesy of the British Astronomical Association*)

although perhaps of a very special kind. This belief was strengthened somewhat when Kraft[12] demonstrated that a fair proportion of the U Geminorum variables are also binaries and the spectrum, orbital period and physical properties of Nova DQ Herculis were found to be very similar to those of a typical dwarf nova. There are differences, naturally. For example, in the same way that the light curves of individual novae vary widely, so also do the spectroscopic details among these stars at minimum. Allowance must be made for the interference of the nebular gases still surrounding the nova (material which is, in all probability, not present in the case of the dwarf novae), but even when this is taken into account, we find a wide diversity of spectral features.

Some novae such as GK Persei (1901) do exhibit composite spectra with the absorption lines of a late-type star in addition to the more usual emission features; others, for example, DQ Herculis (1934), have very strong emission lines, while DI Lacertae (1910) has only narrow, weak emission lines with broad, shallow absorption features. HR Lyrae (1919), on the other hand, has an almost continuous spectrum. A further important distinction between the spectra of the novae and the dwarf novae is the higher excitation of the emission lines found in many old novae. This is most conspicuous in the case of the line of ionised helium at λ 4686 which is seen strongly in emission in most novae at minimum but is rarely visible in the spectra of the dwarf novae. It is, however, seen though very faintly in the spectrum of U Geminorum itself. There is really very little mystery about this observation. Almost certainly, as we would expect from their behaviour during a typical outburst, the blue components responsible for this emission are of a higher surface temperature in the old novae than in the U Geminorum variables.

In our discussion of the binary nature of the dwarf novae we saw how the late-type companion completely fills its lobe of the zero-potential surface, overflowing through the inner Lagrangian point to form a ring or disc of gas which surrounds the blue companion and, as proved by its spectrum, accompanies it in its orbit. All of the evidence we have at present suggests that a similar state of affairs exists for the old novae; indeed, some astronomers now tend to group the novae, recurrent novae and the U Geminorum variables into one general class termed the cataclysmic binaries. As we shall see later there appear to be certain fundamental differences between the dwarf novae and the other two classes.

First, however, let us consider another point of similarity between these three groups of variables which has been pointed out by Kraft[13].

Period-Luminosity Relationship

The discovery that both the old novae and the U Geminorum variables are close binaries represents a decisive turning point in the study of these stars. Not only does it answer several of the problems concerning these peculiar variables; it also raises several additional problems, many of which have yet to be solved. Several lines of attack on these problems are being pursued at present and have already yielded important results. It has now become clear that of the novae and dwarf novae so far investigated, a very high proportion are binary systems consisting of a small blue star (which is of the white dwarf class) and a much larger, cooler companion having a dG or dK spectrum and even in the case of those stars for which there is no direct evidence of duplicity, there is also no evidence against the hypothesis that they too are spectroscopic binaries. Now if this picture is correct it may be possible to find some correlation between the orbital period and the luminosity of the late-type component. Kraft[13] has shown that if the yellow star does overflow its lobe of the zero-potential surface, then provided we ignore any distortion of this surface, the following equation holds

$$\frac{R_r^3}{P^2} = \frac{G}{4\pi^2}\frac{\mathfrak{M}_b}{(1+\mathfrak{M}_r/\mathfrak{M}_b)^2} \tag{1}$$

where R_r is the radius of the late-type star, P is the orbital period and $\mathfrak{M}_r$ and $\mathfrak{M}_b$ are the masses of the red and blue stars respectively. Now the luminosity of the late-type component may be written empirically as

$$L_r = k4\pi R_r^2 \sigma T_e^4 \tag{2}$$

where σ is the Stefan-Boltzmann constant and T_e is a pseudo-effective temperature which is found from the spectral characteristics. If we now substitute (2) into equation (1), we may obtain an expression relating the absolute magnitude of the late-type star to the orbital period and certain other parameters.

$$M_v(\text{red}) = -\frac{10}{3}\log P - \frac{5}{3}\log\frac{\mathfrak{M}_1}{(1+\mathfrak{M}_2/\mathfrak{M}_1)^2} - 10\log\frac{T_e\,(\text{red})}{T_e\,(\cdot)} - 2.5\log k + 73.6 \quad (3)$$

In this equation, the symbols have their usual meanings, the value of k being less than unity, probably because the late-type star is continuously losing mass through the inner surface of its lobe. This rather formidable equation may be simplified a great deal. As we know the masses of the U Geminorum components reasonably accurately (and the equation is equally applicable to both novae and dwarf novae), the second term on the right hand side reduces to approximately 0.8 magnitude and the third term, found similarly from the available spectrograms, has a range close to 2.2 magnitudes. Therefore, averaging out these terms we have

$$M_v(\text{red}) = -\frac{10}{3}\log P - 2.5\log k + 8.6 \quad (4)$$

Here the value of P is in hours whereas in equations (1) and (3) all values are given on the c.g.s. system. Equation (4), of course, is only an approximation and in practice we find that there is an intrinsic uncertainty range of about ± 1.5 magnitudes.

Fig. 18 shows the plot of M_v(red) against log P, giving the positions of some of the novae at minimum and also the mean position of the U Geminorum variables (these have been obtained from the statistical parallaxes found by Kraft[14]). Although the amount of data which is at present available is small, it does appear from Fig. 18 that the fainter the nova is at minimum, the shorter is the orbital period. This relationship also applies to the dwarf novae and as Kraft has indicated, among both the novae and the U Geminorum variables, those of longer orbital period (and therefore having secondary components of higher luminosity) show a tendency for the spectra of both stars to be visible at minimum. As the period decreases among the stars of these classes, the secondary component becomes fainter with respect to the primary. One might argue that the effect is one of obliteration of the absorption lines due to rotation. There are, however, several arguments against this. For example, as the red stars fill their lobes of the zero-potential surface, and if we assume that the rotational period of the red star is synchronised with the orbital period, then the effect must represent a real diminution of the luminosity of the secondary component with a decrease in the period.

Where the novae alone are concerned, we can show by comparing the parameters of DQ Herculis and T Coronae Borealis that long orbital period and high luminosity go hand in hand with the larger mass of the red component of the latter variable. In the case of the dwarf novae, however, such a diverse range in mass is not found although the range of periods is quite small compared with those of the novae (and recurrent novae) and verification of the small differences in mass will be extremely difficult.

Galactic Distribution of Novae and Dwarf Novae

We can readily find, both from the known distribution of the novae in our own galaxy, from their radial velocities, and from work which had been carried out on the novae within other galaxies, notably that by Arp[15] who has studied their distribution in M31, that the vast majority of these stars are members of the galactic disc and not of the galactic halo. This, as we have seen, is not the case for the dwarf novae which are not concentrated along the band of the Milky Way, nor do they show any tendency to cluster in Sagittarius, that is in the direction of the centre of the galaxy. Only about 45 per cent of the dwarf novae lie within 10° of the galactic equator whereas between 75 and 80 per cent of the known novae lie in this region of the sky. We must remember at this point, however, that there are several areas of the heavens which have not been as thoroughly searched for dwarf novae as one would wish and it is possible that future discoveries may change this general picture somewhat.

Then again, there are instances known of novae which seem to belong to the galactic halo. One well-known example is T Scorpii (1860) which flared up very close to the nucleus of the globular cluster M80. Certainly we may argue that since this

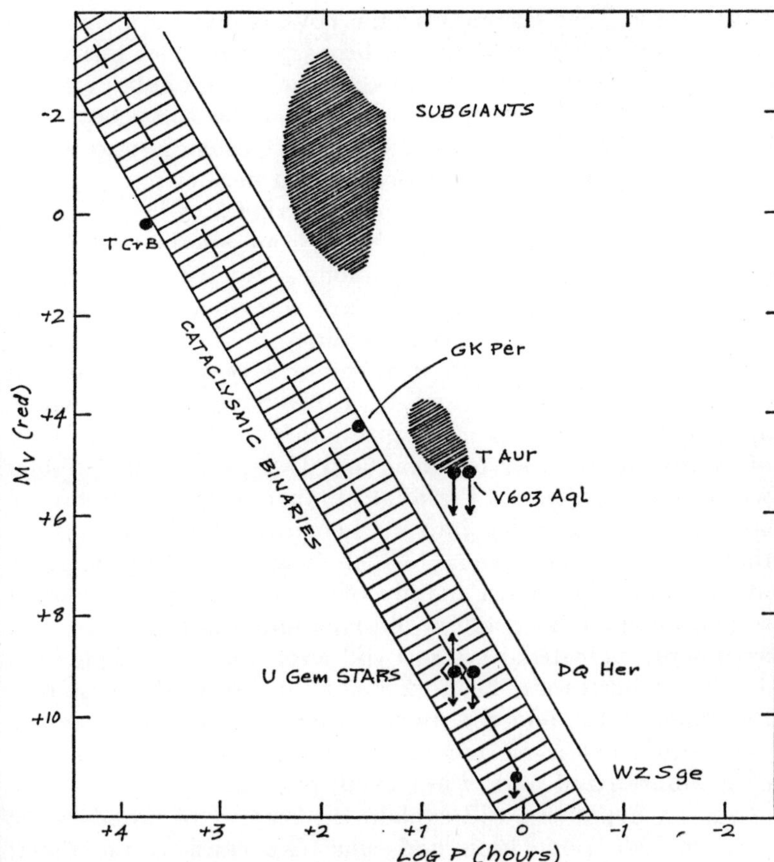

Fig. 18 – The period-luminosity relationship for the secondary components of cataclysmic binaries (*Astrophysical Journal by kind permission of the University of Chicago Press*).

particular cluster is not far from the galactic centre, Nova T Scorpii might be merely a star which is seen projected against the field of M80. There is one fairly convincing argument against this idea, however. Mrs. Hogg[16] has examined all of the available data on this star and from this, coupled with the life-luminosity relationship for novae, has estimated the absolute magnitude at minimum which turns out to be −8.5 magnitude. From equation 1 given in the Introduction, we may therefore obtain the distance modulus for the globular cluster M80 if we assume that the nova really is a member of this cluster and we find this to be 15.7. How does this value compare with the distance moduli for the globular clusters as a whole? The answer is – very well indeed. As Arp[17] has shown, the apparent modulus for these clusters is 15.9 which is so close to that found for T Scorpii that most astronomers are reluctant to dismiss membership of this star with the cluster.

Suppose we attack this problem from another direction. There is now a growing belief among many astronomers that all novae and U Geminorum stars are close binaries. Some even go further and suggest that a condition of a star becoming a nova is that it must be a member of a close binary system consisting of a blue and a red star which have begun anywhere along the main sequence. Accepting this for the moment, then obviously the number of novae and U Geminorum variables will be proportional to the total number of such close binary systems. Without going into mathematical details, it appears that those stellar systems which possess a high angular momentum will consist mainly of stars having a high axial rotation and a binary nature. Equally important to the present argument, such stellar systems will also, because of their high angular momentum, be appreciably flattened (for instance, the galactic disc and the nuclear bulge). Where the angular momentum is low we have the opposite state of affairs; few close binaries and stars of low axial rotation.

Such a hypothesis will explain the scarcity of novae in the galactic halo population and when we consider the dwarf novae, as we have already seen, they have a galactic distribution intermediate between the novae and the halo population.

We now come to one important, apparent difference between the novae and the U Geminorum variables, namely the nature of the outburst. In the case of the dwarf novae, this has already been discussed in some detail in an earlier chapter. There, we saw that there is no evidence in the spectrum of an expanding shell of gas immediately following a typical outburst, nor are there any of the forbidden lines visible in the spectrum which are almost invariably associated with a nova explosion.

To complete the comparison between the two classes of star, we must now examine the nature of the outbursts among the novae. While the sequence of events which follow a nova eruption are readily understood, those leading up to the explosion are still mainly a matter of conjecture and, quite naturally, several theories have been advanced to explain the cause of the outburst. This is readily understandable since after the star has turned nova, there is a profusion of observations available, both photographic and spectroscopic, whereas very little is known of these stars prior to the outburst and in only a small number of cases is the spectrum of the pre-nova stage known in any detail. Nova Delphini (1967) is one star which was sufficiently bright before its outburst for the spectrum to have been photographed, being of type O9. In so many cases, however, we are wise only after the event.

There are, of course, certain facts about which we are reasonably certain. Due to some unstable mechanism within the star (most probably in the outermost layers), a mass of gas which is predominantly hydrogen, is thrown out by the star resulting in a tremendous outflow of energy which appears as both light and heat. The radial velocity of this gas, which may be either in the form of a spherical shell or ejected in two preferred directions (the latter being the most common), is very high and of the order of 800 to 1,200 kilometres per second. After a time it may even be possible to see this gaseous envelope visually; that surrounding Nova Aquilae (1918) being seen by Barnard with the 40-inch Yerkes refractor some six months after the outburst when the nebulosity was only two seconds of arc in diameter. Two years later, its diameter had doubled but by now it was at the limit of visual detection. Long exposure photographs taken in red light by Baade in 1941 showed it fairly clearly and by the use of the same technique, Baade was able to photograph the similar nebulous envelopes around Nova Aurigae (1892), Nova Cygni (1920) and Nova Lacertae (1936).

It follows also that since several recurrent novae are known, this process of ejecting an outer envelope of hydrogen may take place, not once, but several times. In the much rarer case of a supernova explosion, a different sequence of events must be postulated since most, if not all, of the star is disrupted in the process.

Coming now to the various theories concerning the actual

physical nature of nova outbursts, let us first discuss those which are based on the assumption that such stars are binaries, in which case it is now agreed that it is the blue star which is the seat of the explosive process. The positions of these blue components on the Russell-Hertzsprung diagram vary quite considerably. Several are fairly typical white dwarfs. Those of Nova GK Persei (1901) and the recurrent nova T Coronae Borealis are found in the region occupied by stars which are sufficiently evolved to have begun helium burning. Obviously, whatever mechanism we advance it does not depend to any great extent on the mass of this star or on its position on the Russell-Hertzsprung diagram.

One theory which was elaborated by Kraft[12] but which he later rejected is that the blue component is an electron-degenerate star having a very thin non-degenerate atmosphere of hydrogen which is only about 100 kilometres in thickness. If now, some of the hydrogen in the gaseous ring surrounding this star (material which is escaping through the inner Lagrangian point from the red companion) succeeds in penetrating through this shallow atmosphere down into the degenerate core, the internal temperature of the blue star will be sufficiently high to start hydrogen burning. Now one peculiarity of electron-degenerate matter is that unlike normal matter it will not cool by expansion and as a result the rate of reaction will increase very rapidly indeed until the material in the core is no longer degenerate. Expansion of the star will then take place in a very short time, certainly within a matter of hours, possibly even minutes, and the star will become a nova.

In spite of the obvious facts that such a hypothesis satisfies observation insofar as we know there is a ring of gas present and the physical characteristics of the system are almost exactly those required by this idea, some very serious doubts have been raised against this theory. For one thing, we have stars such as T Coronae Borealis in which the blue star appears to be far too massive to be electron-degenerate. Secondly, some more recent calculations show that the conductivity of electrons throughout an electron-degenerate mass is of a very high order and if the temperature is raised at any one point, as it would be if part of the gas stream penetrated a thin atmosphere, the heat will be conducted through the entire mass very efficiently indeed. This means, of course, that before degeneracy can be removed, the entire core of the star must be heated to a high temperature. This, in itself, is not an argument against the theory. The crucial test is – what will happen when the whole core is raised to the temperature necessary for explosion? We now have reason to believe, as Mestel[17] has shown, that once this happens an explosion will certainly occur, but on a truly catastrophic scale and instead of the star becoming a nova, a supernova explosion will result.

Turning to an alternative theory, Cowling[18] demonstrated, in 1941, that if the axial rotation of the white dwarf is not synchronised with that of the other companion it is possible for non-radial oscillations to occur in the blue star. More recently, Cowling's ideas have been further developed by Schatzman[19] who has suggested that, because of the presence of such long period oscillations, material from the blue star will be ejected after a certain induction time which will depend upon the physical characteristics of the system. It is significant that this idea supposes that the gases will be thrown out, not in the form of a spherical shell, but as a cone which again is in good agreement with observation. We must note, however, that on this hypothesis there is no requirement that the binary stars must form a blue and red combination. Any pair of stars will suffice provided that one is of the white dwarf type. There is no need for the companion to fill, or overflow, its lobe of the zero-potential surface.

We must finally discuss a third hypothesis of nova outbursts which does not require the presence of a companion star. To understand the mechanism of the reactions leading up to the actual explosion, we must first consider the way in which energy is produced within a star. Initially, such energy is produced by the conversion of hydrogen into helium by means of a series of reactions known as the proton-proton chain reaction. Basically, this consists of three stages. In the first reaction, two protons (the nuclei of ordinary hydrogen) combine to form a deuteron, a particle whose nucleus consists of a proton and a neutron, a positron being ejected in the process. In the second stage, a further proton is added to the deuteron giving an isotope of helium accompanied by the emission of gamma radiation. Finally, two of these helium nuclei combine to form ordinary helium and two protons which are then available to continue the reaction. The energy from this series of nuclear reactions comes from three sources – the emitted positron, the radiation and the energy of motion of the two ejected protons. A third particle known as a neutrino is also emitted during the first stage of the reaction. This is a very peculiar particle in several respects. It possesses neither mass nor an electrical charge and takes part in no nuclear reactions. Its only function appears to be to carry energy and accordingly the emission of neutrinos represents an energy loss to the star.

Now the most favoured place for the proton-proton reaction to take place is in the centre of the star where both temperature and pressure are highest and gradually, a core of helium is formed surrounded by a thin energy-producing skin in which the above reactions are taking place. There now appears to be a good deal of evidence against the belief, originally held, that any appreciable mixing of hydrogen and helium occurs within the star. As time goes on, the energy-producing shell works its way outward towards the surface with the helium core growing at the expense of the outer hydrogen atmosphere.

We must now consider what happens when this energy-producing layer approaches the stellar surface. This is a vitally important problem since the temperature required for the proton-proton reaction to proceed is somewhere in the neighbourhood of 10 million degrees. Is it possible for the surface temperature of a star to be as high as this? The answer is quite clearly, no; for if this were to happen it would be utterly impossible for the star to radiate such a tremendous amount of energy into space, the surface area is simply not large enough.

In spite of this, it has been found that stars do use up most of their available hydrogen. In other words, this energy-generating shell comes very close to the stellar surface. The problem now is how is the remaining hydrogen consumed? One way would be for convection currents within the outer atmosphere to circulate the hydrogen down through the reaction medium bringing about an interchange of hydrogen and helium. This will almost certainly occur to a certain extent although it is believed that the amount of surplus hydrogen used up in this way is quite small.

The predominant reaction is believed to be explosive in nature, due to most of the remaining hydrogen becoming degenerate. Since this explosive condition occurs at the surface of the star rather than deep in the interior, the hydrogen is thrown off with very high velocities; not high enough, however, for most of this material to escape entirely from the gravitational field of the star. In time, as the heated gas cools, it falls back and returns to its original state so that we may expect the process to occur again.

Each of the above theories has points to commend it although the fact that, of those old novae which are sufficiently bright for high dispersion spectrograms to be obtained, all have shown indications of a binary nature would lend support to the idea that a prerequisite for a star to turn nova is that it should be a binary of a very special kind, namely one consisting of a white dwarf and a much cooler, red star which fills its lobe of the zero-potential surface.

All of this would suggest that whereas there are obvious similarities between the light variations of the dwarf novae and the novae themselves, the similarity becomes much more vague when we consider their galactic distribution and the actual nature of the outbursts. It is certainly true that both classes of star appear to be binary systems, one component of which is a white dwarf and their orbital periods are very alike.

REFERENCES

1 Kraft, R. P., *Ap. J.*, **127**, 626 (1958)
2 Kraft, R. P., *ibid*, **130**, 110 (1959)
3 Walker, M. F., *Publ. A.S.P.*, **66**, 230 (1954)
4 Walker, M. F., *Ap. J.*, **123**, 68 (1956)
5 Walker, M. F., *I.A.U. Symposium* No. 3, Ed. G. H. Herbig (Cambridge, Cambridge University Press) (1957)
6 Walker, M. F., *Ap. J.*, **127**, 319 (1958)
7 Walker, M. F., *ibid*, **134**, 171 (1961)
8 Greenstein, J., *ibid*, **126**, 23 (1957)
9 Greenstein, J., *Stars and Stellar Systems*, Vol. 6, Ed. J. Greenstein, (Chicago, University of Chicago Press) (1960)
10 Wallerstein, G., *Publ. A.S.P.*, **75**, 278 (1963)
11 Sanford, R. F., *Ap. J.*, **109**, 81 (1949)
12 Kraft, R. P., *ibid*, **135**, 408 (1962)
13 Kraft, R. P., *ibid*, **139**, 457 (1964)

14 Kraft, R. P., *Helen B. Warner Lecture of the American Astronomical Society*, April 18, 1963
15 Arp, H. C., *Ap. J.*, **61**, 15 (1956)
16 Hogg, Helen S., *Comm. David Dunlap Observatory*, No. 1 (1938)
17 Arp, H. C., *Stars and Stellar Systems*, Vol. 5, *Galactic Structure*, Ed. A. Blaauw and M. Schmidt (Chicago, University of Chicago Press) (1964)
18 Cowling, T. G., *M.N.*, **101**, 368 (1941)
19 Schatzman, P., *Ann. d'ap.*, **12**, 39 (1958)

8. Relationship to the W Ursae Majoris Variables

In the previous chapter, we reviewed the evidence which has been put forward to support the relationship of the dwarf novae to the ordinary and recurrent novae. There are, as we have seen, several points of similarity between these three classes of cataclysmic variables and from the general appearance of their light curves alone it is perhaps inevitable that the early investigators should have considered them all to be generically related. Such a belief was strengthened by the discovery that all three classes of object are close binary systems in which one component is a very hot, blue subdwarf star very like the white dwarfs and clearly one which is in an advanced state of evolution, almost certainly at the stage of helium burning in its core.

However, not all astronomers share this view, mainly because, as was pointed out earlier, they have different galactic distributions and it now seems fairly conclusive that the physical mechanism of the outbursts is not the same. When we come to discuss in detail the nature of the dwarf nova outburst we shall see that there is now some direct observational evidence that, in the case of U Geminorum at least, it is the cooler, secondary component which is the seat of the eruptions, this being the very opposite of that found for the novae and recurrent novae. Now if we are to disregard the theory that the novae and dwarf novae are related, what do we have to put in its place? This is a question which has gained increasing importance in the study of these stars during recent years and at the present time quite a lot of observational data have accumulated to suggest that there may be an even stronger link between the dwarf novae and a class of variable star known as the W Ursae Majoris variables.

At this point in the discussion it will be appropriate to consider some of the properties of the W Ursae Majoris stars in a little detail before we continue with the important question of the similarities which exist between them and the U Geminorum stars.

The W Ursae Majoris Variables

The prototype of this class of variable was discovered in 1903 by Müller and Kempf and has been fully described by Woodward[1] using all of the observations available up to 1942. This particular star has a photographic magnitude of 8.30 at maximum and exhibits primary and secondary minima of 9.06 and 8.98 magnitude respectively. The light curve (Fig. 19) is obviously typical of an eclipsing system, very similar in many ways to those of the well known β Lyrae variables with no evidence of flat maxima as are found in the related Algol stars, the special characteristic of the light curves of these variables being that they show a continuous variation in brightness and the minima are of almost equal depth. In addition, and this is very important when we come to relate them to the dwarf novae, their orbital periods are all extremely short, far shorter than those of the β Lyrae stars. In general, almost all have orbital periods of less than 12 hours, sometimes even shorter than 5 hours. Once again, therefore, as in the case of the novae and recurrent novae previously discussed, we find orbital periods of the same order as those discovered in the dwarf novae.

These extremely short periods, taken in conjunction with the absence of any flat portion of the light curve at maximum, provide us with another vital piece of information concerning the physical nature of these systems. They are all very close binaries (an alternative name which has often been applied to them is contact binaries for this very reason), the separation of the component stars being so small that they are virtually touching. This latter feature indicates that the stars must both be grossly distorted by the very powerful gravitational and tidal forces exerted by each star upon the other.

When we come to examine their physical characteristics, as deduced from their light curves and spectra, we also find certain differences between them and the much larger class of β Lyrae stars which they closely resemble. The latter variables have components which are both very large and massive with

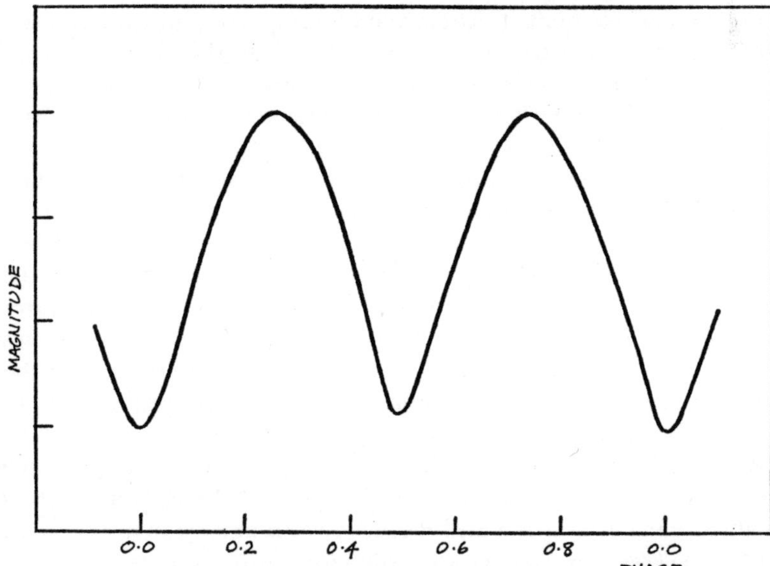

Fig. 19 – Light curve of a typical W Ursae Majoris variable.

diameters often up to fifty times that of the sun, their spectral classes being usually types B or A. The masses of the individual components of the W Ursae Majoris systems, on the other hand are very similar to that of the sun, their sizes are also quite comparable and their spectral types are later, generally of class F or G (the two components of W Ursae Majoris itself are both of class F8p).

Now the light curve shown in Fig. 19 is a mean curve, being the average of several cycles and as a result any small irregularities have inevitably smoothed out in the process. Several observers have studied the light fluctuations of W Ursae Majoris by means of extremely accurate photo-electric determinations covering a large number of individual cycles and this work has revealed certain peculiarities. For example, we find that the relative depths of the primary and secondary minima are not constant and they, together with the brightness at maximum, change over a period, sometimes even from one cycle to the next. Quite clearly, if we have to contend with such short-period variations as these, which are found to be quite unpredictable, it is far from simple to assign any orbital elements to these stars merely from an examination of their photoelectric light curves. Not only this, but the mean periods of these stars show a tendency to vary, that of W Ursae Majoris itself have been increasing very slowly since 1950.

The use of three-colour photometry and the results obtained from it also deserve detailed comment here since they have revealed other curious variations in the behaviour of these variables. Recently, it was found that whereas in 1960 the primary minimum in blue light occurred 0.0001 day later than that in yellow light, two years later the situation was reversed. Such small changes have also been recorded for the secondary minima. At the moment, we are not at all certain of the reason for these anomalies. Some astronomers have suggested that such short-term changes are due to the presence of bright spots on the photospheres of the component stars which vary in position and intensity from one cycle to the next. Others, however, believe that the variations are brought about by the presence of gas streams within these systems and if this is indeed the case (although as we shall see there are now some fairly convincing arguments against such a picture), then here we have a further analogy with the dwarf novae. Just what the effect of the moving gas streams is on the irregular light variations is still a matter for conjecture. It may be that the gas varies in luminosity (and density) from one cycle to the next or it may even absorb some of the light emitted by one or both of the stars, possibly re-radiating it by the process of fluorescence.

The spectra of the W Ursae Majoris variables normally show the narrow absorption lines of both components but the intensities of these lines vary in an extremely puzzling manner. Those lines which are shifted towards the violet end of the spectrum are found to be more intense than the lines which are correspondingly shifted towards the red end and this holds irrespective of which of the two stars is approaching us. A further peculiarity is that during the time of eclipse, the spectral class appears to be later and the absorption lines become single, again regardless of which of the two stars is being eclipsed.

Now we have already said that the masses of the individual stars are appreciably less than those of the β Lyrae variables, being much more comparable with that of the sun. However, although the absorption lines indicate that both components have roughly the same luminosity (and from the mass-luminosity law should therefore be of about the same mass), the radial velocity curves are in direct contradiction with this and show quite clearly that, in general, one component is between two and four times more massive than the other.

This is an extremely important observation in support of the theory that these variables may be an earlier state in the evolutionary path which leads to the dwarf novae since it immediately implies that the more massive component will be the first to begin evolving, moving towards the right of the main sequence as the mass of the helium core begins to increase due to hydrogen burning in the centre, becoming larger in the process. As time goes on, the primary will have expanded to the point where it completely fills its lobe of the equipotential surface as defined by the restricted theory of the three body problem. Any further expansion must now necessarily result in the ejection of mass though either the inner or outer Lagrangian point, theory indicating that it will be predominantly through the inner point when it will be captured by the secondary star. Almost certainly both processes go on simultaneously although to different extents.

Having given the basic characteristics of the W Ursae Majoris variables we are now in a position to consider two important aspects of these stars, namely how they can evolve further into typical dwarf nova systems and equally important, the type of object from which the W Ursae Majoris stars themselves are descended. Since the former is clearly dependent to a certain extent upon the latter, we shall first discuss the possible progenitors of the contact binaries. To do this, we must consider two different models which have been proposed for these variables.

Mass Transfer Model of W Ursae Majoris Systems

Earlier in the chapter, we mentioned how the irregularities in the light variations of these stars might be explained on the basis of a gas stream circulating within the system. Such a model has been postulated for these variables by Sahade[2] and Kraft[3] with the primary expanding to fill its critical lobe of the equipotential surface and as a result ejecting mass through the inner Lagrangian point into the corresponding lobe of the secondary component. This model has been considered further by Kraft[4] who has pointed out that for a star to eventually become the primary of such a system, it must evidently be sufficiently massive for helium formation to have begun in the core during its lifetime, an observation which provides a figure for the lower limit to its mass. From this we find that such a star must have a mass at least 1.25 times that of the sun. Now we already know that the large majority of W Ursae Majoris variables so far investigated contain primaries whose masses are much lower than this (the usual figure is about one solar mass); consequently, we must accept the fact that quite an appreciable mass loss must have occurred during the evolution into the W Ursae Majoris stage.

The two possible modes of mass loss have already been mentioned; transfer to the secondary or lost into space through the first external Lagrangian point. Since we have no observational proof of any mass loss from these systems, not even spectroscopic evidence of any kind, it may be asked if there is any evidence, albeit indirect, which might support this theory. The answer is that some does exist since Kitamura[5] has shown mathematically that where we have a binary system in which one star is gaining mass at the expense of the other, the primary will be underluminous and the secondary overluminous with respect to their masses. Since this is the very result we find for the W Ursae Majoris variables (and also for the U Geminorum stars), it would appear to be a very powerful argument in favour of this particular model.

So far, therefore, there would seem to be a good deal of evidence in support of the present picture of a W Ursae Majoris system and little against. The first indication that the model might prove to be untenable comes from a study of one particular variable of this class – TX Cancri. What makes this variable so important in this context is that it happens to belong to the Praesepe cluster and on the Russell-Hertzsprung diagram for this cluster it lies several magnitudes below the breakoff point, implying that it is on the zero-age main sequence. From this, Kraft[7] has shown that the present theory would indicate that the precursor of TX Cancri was a close, but not a contact, binary in which the mass of the primary was initially about twice that of the sun and the absolute magnitude possibly slightly brighter than 0.0.

Now the mass of the primary has been shown by Popper[6] to be 1.5 solar masses and this star must therefore have lost 0.5 solar masses during its evolution to its present state. During this time, its path on the Russell-Hertzsprung diagram will be downward and roughly parallel with the main sequence. Both Morton[7] and Paczynski[8] have shown that this leads to a mass loss on a Kelvin time scale which is considerably shorter than the age of the cluster itself and consequently the W Ursae Majoris type phenomenon is of only recent occurrence.

The argument against this evolutionary sequence has been detailed by Kippenhahn, Kohl and Weigert[9] who have demonstrated that in such close binary systems, the primary will evolve first of all into a red giant and then into the white dwarf stage and that the time spent in the W Ursae Majoris stage with mass being transferred to the secondary is extremely short compared with the stage preceding loss of mass from the primary. This immediately shows the present model of the W Ursae Majoris systems to be untenable for it would imply that the number of close binaries which satisfy the mass-luminosity relationship for stars capable of evolving into W Ursae Majoris variables would be half of the total number of main sequence stars with spectral classes A to F.

Several attempts have been made to overcome this difficulty but with very little success. In every case, the number of very close binaries consisting of main sequence stars is far too few to explain the large number of W Ursae Majoris stars known.

Stable Stellar Configurations

Since the above model has proved to be unacceptable, certain other progenitors of the W Ursae Majoris stars have been considered, particularly those which evolve on a nuclear, rather than the shorter Kelvin, time scale. One model which satisfies this condition has been described by Lucy[10], this consisting of a pair of stars in contact and surrounded by a common envelope, the outer atmospheric levels of both stars being in convective equilibrium and with the same adiabatic constant. A condition of this model is that, since only those stars of spectral type later than Fo can have outer layers which are in convective equilibrium (those of earlier classes are instead in radiative equilibrium), such binaries cannot exist at zero age and to overcome this difficulty, an early theory of the formation of binary systems has been re-examined, chiefly by Roxburgh[11]. During the early stages of stellar formation, we can visualise a single object which is contracting out of a mass of gas which is composed almost entirely of hydrogen. On the Russell-Hertzsprung diagram, the track of such a star will be vertically downward, moving towards the main sequence. Throughout this contraction process, only convection will be operating within the interior and not until the star is on the point of joining the main sequence will radiation develop as a dominant process in the core.

Now it can be shown that while the process is one of convection, the rotation of the star will increase as it contracts, throwing off mass from the equatorial regions where the momentum is at a maximum. Not until a radiative core begins to form will there be any drastic development in spite of the rapid rotation. At this point however, Roxburgh[11] has shown that the star will be split in two, the mass of the object then being greater than 0.8 solar masses. Now the importance of this particular figure lies in the fact that it agrees exceptionally well with the total mass of the W Ursae Majoris variables which are also 0.8 solar masses or slightly greater.

Huang[12] has suggested the opposite mechanism for the formation of close binaries, namely that they may be derived from wide pairs of stars in which the separation is large compared with their radii. The process by which these wide binaries evolve into more compact systems depends upon the coupling of ionised jets in the systems with the magnetic fields of the stars, the net result of this being a drastic reduction of the orbital angular momentum. This, in turn, leads to a corresponding reduction in their separation and the eventual formation of close binaries.

Evolution of the Dwarf Novae from the W Ursae Majoris Stars

Having discussed the evolutionary processes which may lead to the formation of the W Ursae Majoris variables, we now come to important new concepts which have been put forward within the last decade or so. Already, we have seen that one of the main difficulties which arose when we attempted to relate the dwarf novae with the ordinary and recurrent novae, was that of the widely different galactic distributions of these classes of star. The novae, being predominantly members of the galactic plane are found almost exclusively along, or within, the confines of the Milky Way. The dwarf novae, on the other hand, have small peculiar velocities and belong to a moderately flattened stellar population – the galactic disc. In addition, we also find that they are not members of Baade's Population II stars (those found in the nuclei of the spiral galaxies and the globular clusters – namely those stars which were first formed out of the primal gas of the galaxies). Here then, as we have already seen, we have our first clue that the U Geminorum variables may be closely related to the W Ursae Majoris stars for the latter are the only other known type of binary having short periods which belong to the galactic disc as shown by Struve[13] and Kitamura[5]. We are now in a position, therefore, to examine the many lines of evidence which support such a relationship in more detail.

Spatial Distribution of the Dwarf Novae and W Ursae Majoris Variables

As a first step, let us look at the spatial distributions of these two types of variable. By now, quite a large number of stars belonging to both groups have been identified and their spatial distributions (z = r sin b) have been determined. From results obtained during an initial survey of these stars, Kraft[3] plotted the diagram shown in Fig. 20 using data obtained for 82 U Geminorum and 123 W Ursae Majoris variables. As may be readily seen, there is a very close similarity between their z-distributions. The distance limits for this particular survey were approximately 240 and 320 parsecs for the W Ursae Majoris and dwarf nova variables respectively.

Having now established that there is this very close similarity between the spatial distributions, how do the other physical properties of these variables compare? We have already discussed their mean periods and shown these to be extremely alike (in this context we are, of course, referring to

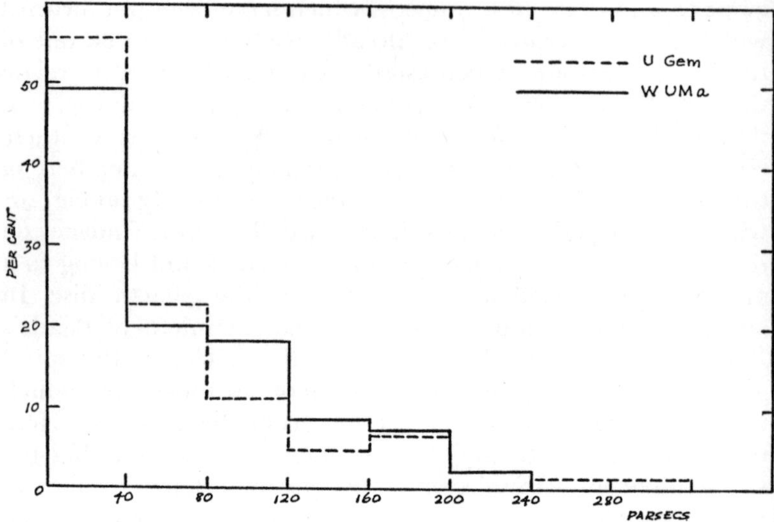

Fig. 20 – Spatial distributions of the W Ursae Majoris variables and the dwarf novae (*Astrophysical Journal by kind permission of the University of Chicago Press*).

the orbital periods of the dwarf novae and not to the far longer and more unpredictable intervals between the major outbursts) and from the data given in Table 4, it will be appreciated that there are several other properties which show a striking resemblance.

Discrepancies in the Luminosities of the Components

Before discussing the important question of whether the above figures can be fitted into an evolutionary scheme in which the dwarf novae develop from the W Ursae Majoris variables, we must first consider further a singular characteristic of both classes of star which has already been briefly mentioned. In each case, one of the components turns out to be seriously underluminous for its spectral class and mass (the other star is overluminous by a somewhat smaller amount but this does not concern us here). In the W Ursae Majoris variables it is the primary which is underluminous by about 3.5 magnitudes and in the dwarf novae, the red, secondary component is also underluminous by a little more, possibly as much as 5 magnitudes. Why this should be so is not known with certainty

Table 4

PHYSICAL PROPERTIES OF U GEMINORUM AND W URSAE MAJORIS STARS

Property	U Gem	W UMa
Mean Period	$0^{d}.25$	$0^{d}.37$
$\mathfrak{M}_1/\mathfrak{M}_1+\mathfrak{M}_2$	0.50	0.33
$\mathfrak{M}_1+\ \mathfrak{M}_2$	1.5–2.0*	1.2–2.5*
Absolute minimum magnitude	+7.5	+4.5
Mean z = (r sin b)	37 parasecs	40 parasecs
Mean peculiar velocity	+16 km/sec	−8 km/sec

*Value given in solar masses

but in the case of the U Geminorum variables at least it would appear to be connected, in part, with loss of mass through the inner Lagrangian point. Kraft[3] has suggested that most of the energy flowing from the star does not go into the luminous flux but rather into raising the potential energy of the gas being ejected through the inner Lagrangian point. There is, however, as he has pointed out, a very serious objection to this point of view. Although we do not know the ejection velocity of this material at all accurately, it is probably somewhere in the region of 20 kilometres per second. From this we may readily calculate the amount of mass lost by the primary of a W Ursae Majoris system assuming the above conditions to hold. A serious problem now arises. If our assumption is correct, we find that the calculated lifetime of a typical W Ursae Majoris system is only between 500 and 1,000 years. A lifetime as short as this cannot possibly be correct for two reasons. Firstly, it would mean that the orbital period would show quite large changes over a relatively short period of time, far larger than any which have been observed and secondly, the total number of such variables would be excessively large. All of this would add confirmation to the view we took earlier when the mass transfer model for the W Ursae Majoris systems had to be abandoned. It seems very likely, therefore, that in these stars some other process is operative, possibly one associated with a change in the gravitational flux due to the high rotation of the star.

Let us now give our attention to the total masses of these two distinct binary systems. If our proposed evolutionary scheme beyond the W Ursae Majoris stage is correct, then the total mass must remain either constant or decrease slightly. This is certainly what we would expect from a W Ursae Majoris system in which hydrogen exhaustion in the core of the main sequence primary results in evolution away from the main sequence with the equipotential surface expanding it until it passes through the first external Lagrangian point. Once this happens, further swelling will result in mass being lost to the system through this point, a process which will bring about a decrease in the orbital period as demonstrated by Kupper[14] and Kruszewski[15]. Since this mass loss proceeds on a Kelvin time scale, the period of this particular phase of the evolution will be short compared with the previous lifetime of the primary on the main sequence.

There is one apparent anomaly which we must consider here. As we have seen, the period of W Ursae Majoris itself has been slowly *increasing* since 1950. Huang[16] has gone into the problem of mass transfer and its effect on orbital periods in considerable detail although admittedly his results have been mainly derived from a study of the β Lyrae binary systems and may not be applicable in the case of such close binaries as those we are now discussing. However, it would seem that if the fraction of mass being lost is small, then the orbital period will *increase*, not decrease, and it may be that at the present time, this condition holds for this particular variable. We must bear in mind, too, that the period of 18 years or so over which this very small increase in period has been noted is very minute compared with that postulated for evolution into a dwarf nova system.

Taking a typical binary of the W Ursae Majoris type, Kraft[3] has suggested that if approximately 0.4 solar mass is lost into space and 0.2 solar mass is transferred to the secondary, the physical characteristics, including the orbital period, of the resulting system are very similar to those found for SS Cygni. If instead, we assume that 0.3 solar mass is transferred to the secondary and none is lost into space through the first external Lagrangian point, the figures we obtain are a reasonably close fit for U Geminorum. Clearly, such an evolutionary scheme has much to commend it but it would be imprudent to suggest that this is the final answer. There are still many serious points which remain to be cleared up. The actual mechanism whereby the primary component of a W Ursae Majoris system may lose anything up to a quarter of its mass during its evolution into the white dwarf stage is not yet fully understood, nor do we know why it is that among the many hundreds of close binaries known at present, the nova-like outbursts typical of the dwarf novae occur only when the mass ratio is close to unity. Regarding this last point, we shall see when discussing the nature of the outbursts that one condition would seem to be that the secondary must be a star filling its lobe of the equipotential surface and which also possesses a convective subsurface layer.

A further grave objection to this theory has recently arisen following the discovery of W Ursae Majoris variables in the intermediate age galactic clusters NGC 188 and M67. From a comprehensive study of these stars, Kurochkin and Kukarkin[17] have shown that the age of these variables apparently coincides. on average, with that of the corresponding stars which lie on the main sequences of these two clusters, this being of the order of several thousand million years. Quite clearly, if this is so, it will infer that the U Geminorum variables are certainly older than this. On this basis, Kurochkin and Kukarkin have suggested that the peculiar behaviour of the spectral lines, in particular those which vary in position with phase, is due to changes in the outer parts of the stellar atmospheres which do not significantly influence the evolution of either component.

The initial results obtained by Kraft[3] on the similarity between the spatial distributions of the U Geminorum and W Ursae Majoris variables have also been criticised recently by Popov[18] on the grounds that whereas the dwarf novae are distributed almost uniformly when viewed in projection out to distances of about 1,300 light years, this is not the case for the W Ursae Majoris stars. These latter variables show a distinct tendency to agglomerate in the vicinity of the sun and also show a marked cut-off in numbers at a distance of about 450 light years in the direction of the galactic anticentre. This cut-off has been explained by Popov on the basis that here we have the edge of one of the spiral arms of the galaxy (that in which our sun is embedded). Kraft[19], on the other hand, regards this cut-off as due to observational selection. The latter view is supported by the discovery of other small agglomerations of these variables at some distance from the sun. In addition, as he has pointed out, most of the patrol surveys which have resulted in the finding of the W Ursae Majoris variables have been those intended for the discovery of RR Lyrae stars and have been made in the direction of the galactic centre. The reason for this is quite straight-forward. The amplitudes of these variables are all quite small, seldom more than one magnitude, and it is only natural that most have been discovered on plates specially prepared for the detection of RR Lyrae variables (stars which have similar small amplitudes and periods but which are not binaries, but pulsating stars).

A reappraisal of the earlier results has therefore been made recently by Kraft[19] using more accurate data which shows that the preliminary proper motions derived for the W Ursae Majoris and U Geminorum variables[3] were seriously in error. The new observational data, which includes all of those variables of both types lying within a cylinder of radius approximately 1,500 light years and height approximately 3,000 light years, placed at right angles to the plane of the galaxy with the axis passing through the sun and its centre on the galactic plane, shows fairly conclusively that not only are the spatial distributions virtually identical but also the kinematics of these stars. The solar motion of the W Ursae Majoris variables as found by Artiukhina[20] and that of the dwarf novae obtained by Kraft and Luyten[21] are the same within the limits of observational error.

The outcome of the present discussion, therefore, is that most of the observational evidence at present available agrees with the hypothesis that the W Ursae Majoris and U Geminorum variables are very closely related. We have, however, been forced to modify the earlier theories of the precursors of the former stars appreciably. There are still problems awaiting solution which is what we might expect considering the great complexity of these systems.

REFERENCES

1 Woodward, E. J., *H.C.O. Circular*, **446** (1942)
2 Sahade, J., *Liége Symposium: Modéles d'étoiles et évolution stellaire*, p. 76 (1959)
3 Kraft, R. P., *Ap. J.*, **135**, 408 (1962)
4 Kraft, R. P., *Publ. A.S.P.*, **79**, No. 470 395 (1967)
5 Kitamura, M., *Publ. Astr. Soc. Japan*, **11**, 216 (1960)
6 Popper, D. M., *Ap. J.*, **108**, 490 (1948)
7 Morton, D. C., *ibid*, **132**, 146 (1960)
8 Paczynski, B., *Acta. Astr.*, **15**, 89 (1965)
9 Kippenhahn, R., Kohl, K., and Weigert, A., *Zs.f. Ap.*, **66**, 58 (1967)
10 Lucy, L. B., *Ap. J.*, **144**, 466 (1967)
11 Roxburgh, I. W., *ibid*, **143**, 111 (1966)
12 Huang, S-S., *Ann. d'Ap.*, **29**, 331 (1966)
13 Struve, O., *Stellar Evolution* (Princeton: Princeton Univ. Press), p. 183 (1950)
14 Kuiper, G. P., *Ap. J.*, **93**, 133 (1941)
15 Kruszewski, A., *Advances in Astronomy and Astrophysics*, Vol. 4, Ed. Z. Kopal (Academic Press, London), p. 233 (1966)
16 Huang, S-S., *Ap. J.*, **61**, 49 (1956)
17 Kurochkin, N. E., and Kukarkin, B. V., *Astron. Zh.*, **43**, No. 1, 83 (1966)
18 Popov, M. V., *Peremennye Zvezdy*, **15**, 115 (1964)
19 Kraft, R. P., *Ap. J.*, **142**, 1588 (1965)
20 Artiukhina, N. M., *Peremennye Zvezdy*, **15**, 127 (1964)
21 Kraft, R. P., and Luyten, W., *Ap. J.*, **142**, 1041 (1965)

9. The Dwarf Nova Outburst

Our knowledge of the nature of the outbursts of these variables is comparatively recent. Even now, there are several problems still awaiting solution as the reader will have become aware from the preceding chapters. The old idea that the sudden increase in brightness can be explained in a manner similar to that advanced for the novae and recurrent novae is no longer acceptable following the discovery that the spectroscope shows no evidence of any expanding envelope of gas during the time of a maximum. Quite obviously, therefore, we must look much deeper into the physical nature of these systems for an explanation of the outbursts.

As we have seen in Chapter Four, all of the available evidence suggests that the dwarf novae are very complex binary systems and clearly there are several regions within such systems which may be responsible for the outburst. Until very recently, all of the theoretical considerations have generally assumed *a priori* that the cooler secondary component may be disregarded since this is a yellow dwarf similar in many ways to our own sun. The one marked peculiarity of this star is, of course, that it appears to be grossly underluminous for its spectral type and mass but as we have seen, this may be explained on the basis of the mass loss from this star which is a consequence of the evolutionary path followed by such close binaries. In addition, since these variables show many similarities to the novae, there has naturally been a tendency to regard the hot, blue star, or some region in close proximity to it, as the seat of the nova-like eruptions. That this is an argument with considerable force is clear, but later in the chapter we shall see that in the case of U Geminorum, at least, there is now very strong observational evidence to show that the eruptions in this particular system occur in the atmosphere of the cooler component. The crucial question of whether the same situation holds for the other dwarf novae is one which cannot be answered at present and must be held over until further data accumulate. Unfortunately, as will become apparent during the detailed discussion of the results which have led to the above conclusion, only those dwarf novae which exhibit eclipses in their light curves provide us with this vital piece of evidence and at the moment very few of these are known.

Before considering this important contribution to our knowledge of these stars, let us first examine the other regions of these binary systems. Beginning first with the blue star, we have seen that the masses of these white dwarfs lie very probably in the range 0.9 to 1.2 solar masses. This is within the mass range of normal white dwarfs for which the theoretical upper limit is 1.2 times that of the sun and it is probably significant that the blue components of the recurrent novae are higher than this, probably as high as 3 solar masses (for example T Coronae Borealis). The greater mass of the blue stars in the recurrent novae systems indicates that these objects are far from typical white dwarfs and may be the reason for the explosive outbursts of these variables which have been conclusively shown to be similar to those of the novae themselves with an ejection of mass from the hot component in the form of a symmetrical (or possibly asymmetrical) shell.

When we were discussing the various theories concerning the nova outburst, we saw how it is necessary to explain an eruptive form of outburst to fit the visual and spectroscopic observation that mass is ejected from the nova system either as a spherically expanding shell or in two preferred directions as cones of hydrogen-rich gas. Here there is no difficulty in ascribing the seat of the outburst to the blue component. Since the spectroscope shows no indication of such a mass of outward-moving gas during a dwarf nova outburst, such hypotheses are untenable when we come to examine the maxima of the U Geminorum and Z Camelopardalis variables. Can we therefore explain the abrupt increases in brightness of between four and six magnitudes on the basis of an equally abrupt rise in the surface temperature of the blue star without postulating that there is any appreciable change in the overall dimensions of the star during the course of an outburst?

Now we have already seen that if we assume these white

dwarfs radiate essentially as black bodies at maximum (an assumption which, since the spectrum at this phase is continous with only feeble emission and absorption lines of hydrogen and singly ionised helium visible, is probably not too far from the truth), the surface temperatures at maximum lie between 12,000 and 15,000°K. The validity of the present argument therefore depends very strongly upon the surface temperature of the white dwarf stars when at minimum. If this turns out to be significantly lower than when at maximum we have a strong point in favour of a predominantly temperature effect and do not have to postulate an eruptive outburst of the kind found in the novae.

Unfortunately it is not an easy matter to make accurate direct estimates of the surface temperatures of the blue companions when at minimum brightness since at this phase the light of the cooler component exerts its maximum effect. Now it will be recalled that in certain of the dwarf novae, notably RU Pegasi and SU Ursae Majoris, the blue star is relatively fainter at minimum than the late-type secondary by up to two magnitudes. If there is nothing abnormal about the yellow dwarf star *at minimum*, we may assume a surface temperature in the region of 5,500°K for this component, similar to that of our own sun (when we come to discuss the case of U Geminorum in detail, we shall see that this secondary component has a far higher surface temperature than this at maximum phase).

In estimating the surface temperature of the blue star at minimum, we must be careful to take into account the very different radii of the two components. In Chapter Five, the spectral types of the yellow components of SS Cygni and RU Pegasi were shown to be dG5 and dG8 respectively, with radii of 6.2×10^{10} and 5.9×10^{10} centimetres. For the purpose of the present argument we may assume a mean radius of 6.0×10^{10} centimetres or 600,000 kilometres.

Let us now give our attention to the white dwarf stars in these systems. Since they are probably fairly typical of the class as a whole, we may obtain an approximate estimate of their radii from a study of such stars as the Companion of Sirius which has been shown to have a radius about twice that of the Earth, namely about 13,000 kilometres. As the areas of the radiating discs are proportional to the squares of their radii, we find that the area of the secondary star's disc is about 2,000 times that of the primary. From this, we must therefore conclude that the surface temperature of the white dwarf when the variable is at minimum brightness does not differ appreciably from that when at maximum, otherwise we would expect the primary to be fainter at minimum phase than is actually observed. Some support for this comes from an examination of the position of the white dwarfs in normal systems such as that of Sirius and Procyon on the Russell-Hertzsprung diagram. Here we find them below the main sequence in the region of stars with surface temperatures between 9,000 and 12,000°K.

So far in this discussion we have taken the white dwarf components of the dwarf novae as being typical of such stars in general. How far are we justified in making this assumption? Obviously, we must modify this statement slightly. Being members of close binary systems, we might expect there to be some difference in their surface temperatures, particularly because of the presence of the rotating ring of gas surrounding these stars. In the case of our own sun, there is evidence that as it moves through space it sweeps up some of the tenuous interstellar gas by virtue of its gravitational field, these gases falling in towards the sun at velocities of up to 600 kilometres per second and producing what is known as the splash corona. In the case of the blue components of a dwarf nova system we have a similar state of affairs, but here the density of the gas is far higher and the effect produced by the infalling material correspondingly greater. The net result will be to increase the surface temperature of these stars by about 3,000°K above that of a white dwarf star in a system such as that of Sirius where there is no evidence for such a gas stream. The question of whether the entire surface of the blue star is uniformly increased in temperature or only a portion of it is one which must be considered and we have evidence from the work of Mumford[1] and Krzeminski[2] that there is a hot spot on the following hemisphere of the blue star. The further evidence which indicates that the blue star is not concerned in the outbursts will be discussed in detail later in the chapter. At present it will be sufficient to state here that any increase in the brightness of the blue component during a maximum contributes only about ten per cent of the total light of the system during this phrase.

Let us now go on to consider the gas stream as a possible cause of the abrupt increases in brightness of these variables. There are three probable ways in which this gas may bring about an outburst such as is observed. If the temperature of this gas were suddenly increased by a few thousand degrees, the output of visible light could almost certainly produce a maximum of the kind found in the light curves of the U Geminorum and Z Camelopardalis stars. Unfortunately, at present, we know of no mechanism by which such a large and rapid increase in the temperature can be brought about unless the proportion of radiation in the ultra-violet from the white dwarf becomes markedly increased and there is no evidence of this.

Suppose we now consider the effect of eclipse of the gas stream by the secondary component. Would this produce the desired effect? This idea also has to face up to several criticisms. Firstly, it can be shown that although, in certain of the dwarf novae, the blue companion is totally eclipsed by the secondary, it is unlikely that any eclipse of the gas stream is more than partial. This being so, any diminution in the total light of the system due to eclipse of the stream of gas will be quite small, certainly not more than one magnitude even in the most favourable case. Not only this, but the form of the light curve would be utterly different from that which we find for the dwarf novae, resembling those of the Algol or β Lyrae eclipsing variables.

The third suggestion is that the gas stream, which is composed mainly of hydrogen at quite a high temperature, may actually penetrate the outer layers of non-degenerate matter around the blue companion. Such a condition has already been discussed in Chapter Six when we considered the outbursts of the novae and we saw there that the reaction of this gas with the degenerate material of the core would result in a supernova explosion owing to the highly peculiar properties of degenerate matter. Evidently, therefore, we must reluctantly dismiss the gas stream as a cause of the dwarf nova outbursts.

So far, we have said little about the role of the cooler

component of these systems except for indicating briefly that recent evidence would suggest that it is this star which is the seat of the nova-like outbursts. Such a result goes against all previous theories of the dwarf nova phenomenon and to appreciate the importance of this work, which has been carried out by Krzeminski[2], it is necessary for us to go into it in some detail. So far, the only variable for which it has been shown that the secondary component is the one responsible for the outbursts is U Geminorum but, as will be discussed in the following chapter, it may well be that this important discovery may be adapted to explain the curious 'standstills' of the small subgroup of Z Camelopardalis stars.

Using the 21- and 42-inch telescopes of the Lowell Observatory, the 36-inch Crossley instrument at the Lick Observatory and the 69-inch Perkins reflector at Flagstaff, Krzeminski has obtained a long series of photoelectric measurements either in yellow or blue light, or in three colours; yellow, blue and ultraviolet with a refrigerated 1P21 photomultiplier, from December 1961 to January 1963, to examine the shape and depth of the primary minimum of the star during eclipse. Throughout this period, there were five maxima of the variable of which four were well covered by the observations and it is therefore possible to determine those changes occurring in the primary minima throughout the major light cycle of the star and in particular whether there is any change in the curious shoulder which precedes the descent to minimum and is well shown in Fig. 10. We saw there how Mumford had found marked dissimilarities in the shape and depth of the primary minimum which were very difficult to explain. We also saw that at the time these observations were made, U Geminorum was beginning a rise towards maximum which suggests that perhaps the fact the brightness of the star was also changing appreciably due to the rise, may have had an effect on the minima due to the eclipse.

The photoelectric measurements as determined by Krzeminski indicate that the light curve is very similar to that of VV Puppis which is also an eclipsing variable (although not a dwarf nova) and is, indeed, the only star known at present with a light curve even approximating that of U Geminorum during eclipse. Consequently, the peculiarities of the primary minimum of the dwarf nova which are of particular interest as far as Krzeminksi's argument is concerned, are the smoothly rounded shoulder which precedes the descending branch of the light curve having its maximum at phase 0.85 period, the flat base of the minimum which shows quite conclusively that the eclipse is total, the absence of any secondary minimum and the fact that the descending branch consists of two separate segments, the lower one, just before totality, often showing quite pronounced distortions from one cycle to the next.

In general, the minima show very little difference during the period when the variable is at minimum. During a rise to an outburst, however, the minima become progressively shallower although still retaining their general shape, showing that the eclipse is still total. When the dwarf nova is at its maximum brightness, no eclipses were detected although the very small, irregular fluctuations which are always present, continue. As the star fades, the minimum due to the eclipse once more become apparent, the depth of the eclipse increasing steadily as the star approaches minimum. Krzeminski has also been able to show that even during the course of an outburst of five or six magnitudes, the brightness of the object which is being eclipsed does not increase by more than a factor of five. Not only this, but the region within the system which produces the shoulder on the descending branch of the curve shows no more than a two-fold increase in brightness.

Certain other parameters of the light curve during primary eclipse also change throughout the 102-day cycle of U Geminorum and have a direct bearing upon the model proposed by Krzeminski for this system. The width of the eclipse, as measured at half-depth, has a maximum of about one-tenth of the orbital period when the variable is near the peak of an outburst and then shows an exponential decrease to about one-twentieth of the orbital period shortly before the onset of the succeeding eruption. The eclipses are also found to be wider when measured in the ultraviolet than in the yellow or blue.

The egress of the eclipse, as measured by the duration and slope of the rising branch is also closely related to the main light cycle. This is not, unfortunately, an easy parameter to measure accurately since the period of egress lasts less than 3 minutes. In spite of this difficulty, Krzeminski has shown that the time of egress is longest just after an outburst, shortens by about half before the next major maximum and possibly increases during the beginning of an eruption.

Finally, the observed time of an eclipse is not constant but occurs earlier than calculated when the variable is rising to a maximum at the time of an outburst and then returns to its normal phase as the brightness of the dwarf nova declines following an eruption.

We are now in a position to discuss the physical nature of U Geminorum, taking into account all of the observational data just mentioned. The model proposed by Krzeminski is basically similar to that which has been described earlier in Chapter Four consisting of a hot, blue primary and a larger, red companion with a mass ratio close to unity; the cooler companion filling its lobe of the equipotential surface and being the slightly more massive of the two. Since the secondary star fills its Roche limit, there is an ejection of mass through the inner Lagrangian point forming a rotating ring of hydrogen-rich gas around the white dwarf star. The luminosity of this ring is not, however, uniform but brighter in the region of the inner Lagrangian point, this bright region extending around to the following hemisphere of the white dwarf star.

This non-uniformity of brightness is also apparent in the blue star itself. This is shown from the fact that the region responsible for the rounded shoulder in the light curve preceding the descending branch of the eclipse is also eclipsed. This being the case, the source cannot be connected with the secondary star. Evidently it is either associated with the ring of gas or the blue star. Now we can differentiate between these two possibilities since fortunately the light curve provides us with the necessary clue. The shoulder, as shown in Fig. 10, extends over approximately half of the orbital period and it must therefore be associated with the blue star and not the ring. The most likely explanation is that there is a hot spot on the surface of the primary as suggested by Mumford[3].

We shall now examine the hypothesis put forward by Krzeminski[2] that it is the red component which is directly responsible for the outbursts. We have already seen that the shape of the primary minimum due to the eclipse remains virtually constant on both ascending and descending branches of the outburst although the depth of the eclipse diminishes during the rise and increases once more on the return to a normal minimum. From this observation, we may conclude that the total light contributed to the system by the object which is being eclipsed, that is the white dwarf star and the bright portion of the rotating ring of gas, also remains essentially constant throughout the course of a typical outburst; certainly it does not show more than a five-fold increase if the eclipse remains total throughout the whole light cycle, or ten-fold if we assume that for some reason there is only a partial eclipse during an eruption. However, when passing through a bright maximum with an amplitude of approximately six magnitudes, there is a hundred-fold increase in the total visible light from the variable. From this, Krzeminski has been led to the conclusion that it is the red secondary component which must contribute the remaining ninety per cent of the light.

This very important conclusion leads us on to consider the actual nature of the outburst. In Chapter Three, mention was made of the fact that only two of the U Geminorum variables show the spectrum of the secondary star in the photographic region, these being SS Cygni and RU Pegasi. In the case of U Geminorum itself, the absorption spectrum of the cool star is not visible, indicating that when the variable is at minimum brightness, the secondary contributes only a very small percentage of the total light of the system. When we come to consider the situation at maximum brightness, we find that the effective temperature of the system (obtained by assuming that it radiates essentially as a black body), is about 15,000°K. Since, at this phase, the secondary contributes almost all of the light of the system, its surface temperature must be very nearly 15,000°K. When at minimum, however, its surface temperature is only about 5,500°K, this figure being in good agreement with the spectral types of the secondaries in those dwarf nova systems where both spectra are visible, namely dG5 to dK0.

In discussing the behaviour of the secondary during a typical outburst, we must of course, also consider any change in the radius of this component. As in the case of other eclipsing variables, such as the Algol and β Lyrae stars, we may do this by an examination of the width of the primary minimum due to the eclipse. As has been mentioned earlier, the photoelectric observations show that there is an approximately two-fold increase in the eclipse width during an outburst, from which we may assume that the effective radius of the secondary also increases at this time. To obtain an accurate estimate of how much the radius increases we would, of course, need to know accurately the orbital inclination of this system. From previous considerations, we know this to be fairly close to 90°. If the inclination is exactly 90°, then the secondary will double its radius during an eruption. However, this is not believed to be the case for U Geminorum, the true orbital inclination being slightly less than this. From the dynamics of such an eclipsing system, we may put forward two alternatives. If the secondary just fills its lobe of the equipotential surface when the variable is at maximum brightness, the orbital inclination turns out to be 81° and the radius of the secondary increases by a factor of 1.6 during eruption. On the other hand, if the secondary fills its lobe when the dwarf nova is at minimum brightness, the inclination of the orbit is only 71° and the secondary's radius is increased by only a factor of 1.2. Since it would appear that there is an outflow of matter from the secondary, through the inner Lagrangian point and into the region surrounding the blue star, the situation present in the U Geminorum system would seem to lie somewhere between these two extremes and the increase in the radius of the secondary component is perhaps best represented by a factor of 1.4.

The picture we now have of the U Geminorum system, therefore, is one where the cooler star increases in size by a relatively small amount, forcing mass through the inner Lagrangian point at the beginning of an outburst as a consequence of which we find the brighter part of the ring between the inner Lagrangian point and the following hemisphere of the blue star. At the same time, this ejection of mass from the upper atmospheric levels of the secondary component reveals the lower, hotter levels, resulting in a very marked, and sharp, increase in the surface temperature.

One final point still remains to be explained, namely the

anomalous phase shifts of the eclipses during the rise to an outburst. As found by Krzeminski, the eclipses occur earlier than predicted when the light of the variable is increasing to a maximum, the most likely explanation being that the star does not expand symmetrically, the radius increasing asymmetrically in the direction of its orbital motion.

Finally, of course, there is the question of why the red star should undergo these irregular expansions and loss of mass. At present, we have several theories which are under active consideration. The most likely possibility is that derived from recent work by Paczynski[4] who has shown that the red stars of the dwarf nova systems are all similar in one important respect, namely that they possess convective subsurface zones and in those cases where mass loss occurs, such stars show an instability on the pulsation time scale, unlike those which are in thermal equilibrium where any instability is on the longer Kelvin time scale. As a result, there will be irregular oscillations set up in the red star which are semi-periodic over a few hours and these may be sufficient to trigger off the pulsation in the outer levels which lead to a greater ejection of mass producing the much larger eruptions of these stars. The difficulties associated with computing the actual amount of mass lost through the inner Lagrangian point are extremely great and much more information than is at present available will be required before any accurate estimates may be made.

REFERENCES

1 Mumford, G. S., *Ap. J.*, **139**, 476 (1964)
2 Krzeminski, W., *ibid*, **142**, 1051 (1965)
3 Mumford, G. S., *Publ. A.S.P.*, **79**, 283 (1967)
4 Paczynski, B., *Acta Astr.*, **15**, 89 (1965)

10. The Standstills of the Z Camelopardalis Variables

In the whole of the previous discussion we have made only a brief mention of one of the great mysteries of the dwarf novae, namely the curious 'standstills' which occur in the light curves of the small subgroup of stars named after Z Camelopardalis. It will be recalled that as far as the more normal outbursts of these variables are concerned, it was originally thought that most of their characteristics may be explained on the basis of either an interaction of the stream of high temperature hydrogen-rich gas within the system with the degenerate matter in the core of the white dwarf component or in a change in the opacity of a tenuous medium surrounding this star[1,2]. Now we must tackle the outstanding problem of the 'standstills' using the information gained in the past few years, and in particular why they are found in only a very small number of the dwarf novae and not in others.

Mention has been made elsewhere of the difficulties associated with the classification of the fainter dwarf novae into the two classes of U Geminorum and Z Camelopardalis variables due to lack of observations. Only when a 'standstill' has been unambiguously shown to occur, is it possible to assign a variable to the latter category.

As we know, the 'standstills' are quite unpredictable, both in their occurrence and the period over which they last. Yet in spite of this, they do show some regularities. For example, in the case of any particular star, we find that the onset of a 'standstill' takes place at almost the same point on the descending branch of the light curve, although there are, of course, fairly wide deviations from one variable to another. For instance, in the case of RX Andromedae this point is only slightly below a normal maximum; for Z Camelopardalis itself and TZ Persei, it is about one third of the way down from maximum to minimum brightness. The body of basic observational data gathered over many decades (admittedly for only a small proportion of these variables), also makes it possible to list other generalities which distinguish the Z Camelopardalis variables from those of the main U Geminorum class.

The amplitudes of the Z Camelopardalis variables are, in the main, smaller than those of the other dwarf novae, their periods are shorter and their stay at minimum light a smaller fraction of the mean period. We must also take into account the discovery by Kraft[3] that the radial velocity curve of RX Andromedae, and possibly of Z Camelopardalis, is asymmetrical, unlike those of the U Geminorum variables which have been examined. How far this represents an ellipticity of the orbit is debatable at the present time. Although many astronomers subscribe to this idea, it is just possible that the asymmetry may be due to internal motions of the gas stream.

From a study of the light curves of the Z Camelopardalis stars made over a great many years, certain other curious facts have come to light. With only two exceptions, a 'standstill' has always ended in a decline to a normal minimum, following which the more usual behaviour begins once more. Early in 1959[4], Z Camelopardalis rose from a 'standstill' to a maximum and as recently as September, 1968, TZ Persei[5] has repeated this anomalous behaviour. There is also some evidence, admittedly based upon a small number of visual observations, that in December, 1968, TZ Persei rose from a fairly normal minimum (which actually ended a long 'standstill'), to a second 'standstill' with no intervening maximum[5]. One further point which may be of some significance in a discussion on the nature of the peculiar 'standstills' is that the longest period of comparative inactivity, lasting for approximately eighteen months, occurred shortly after the brightest maximum ever observed for this variable[6].

These then, together with the prolonged periods of completely erratic light variations which are also characteristic of the Z Camelopardalis variables, are the facts concerning the light fluctuations which we have to go on. How far do they go towards providing us with any clue as to the physical nature of the 'standstills' and of the erratic variations? On the basis of the observational data, not very far. From the last point mentioned above, it might appear that there may be some correlation between the peak brightness of the maximum immediately preceding a 'standstill' and the duration of the 'standstill' itself. In this way it would be possible to explain a standstill' as a residual effect of the outburst itself.

When we come to examine the position critically, however, we find very little real evidence of any such simple relationship. Certainly there is the one instance mentioned earlier in respect of TZ Persei but in most other cases a long 'standstill' has followed a quite faint maximum and *vice versa*. Even if we integrate the area beneath the light curve for the relevant maximum, thereby obtaining some measure of the total visible radiation emitted, we find a very similar result. Unfortunately, at the present time, we do not have complete three-colour curves of these variables from which we can measure the total radiation emitted during an outburst which precedes a 'standstill' in order to make a direct comparison with the duration of the succeeding period of inactivity and consequently it would be unwise to be dogmatic on this point. All of the available evidence, however, indicates that since we are reluctantly forced to abandon the idea of such a straightforward relationship, we must probe deeper for some other factor present in the Z Camelopardalis variables which is absent in the more abundant U Geminorum stars.

Our first step will be to consider the three individual regions within the system of a dwarf nova, as we did in the last chapter when discussing the nature of the outbursts, and see whether it is possible to uncover some such factor by a process of elimination. As we have seen earlier, these three regions are – the white dwarf component, the gas stream flowing from the secondary star through the inner Lagrangian point and taking up an orbit around the blue star and the cooler, secondary companion. Obviously, in a closely-knit system such as this all three regions are closely interrelated and any change in one will exert a corresponding effect upon the others and in our present state of knowledge regarding the 'standstills', it is often extremely difficult to assess the magnitude of these changes or indeed what they will be.

Much of the data we have at present suggests that, contrary to the theories which had been advanced prior to 1965[1,2], variations in the radius and surface temperature of the cooler, secondary component are responsible for the nova-like outbursts. As far as the 'standstills' are concerned, the position is far more vague and difficult to assess. From the light curves of the Z Camelopardalis variables given in Part Two, we find that throughout the entire cycle of a 'standstill', in spite of the small-scale fluctuations which occur, the star never attains a brightness comparable with that of a normal maximum (in the case of RX Andromedae where 'standstills' occur at a point only slightly below normal maximum this statement may require qualification). One tentative conclusion we may draw from this is that for some reason no major outburst of the variable occurs during the time of a 'standstill'. This is all the more surprising when we consider that such periods of comparative inactivity can last for a year or more, during which time (since the mean cycles of the Z Camelopardalis stars are, in general, even shorter than those of the U Geminorum variables), we would expect a score or more maxima to take place under more normal conditions. All of this would tend to suggest that in some way, the onset of a 'standstill' suppresses the activity of the mechanism causing the major outbursts of these stars.

Why this should happen is not known with certainty. The discovery by Krzeminski[7] that in the U Geminorum system, it is the cooler, secondary component which is the seat of the nova-like eruptions, now throws a new light on the problem of the 'standstills'. We must, of course, bear in mind that it is only in the case of this one variable that we have this information from the accurate photoelectric study of the eclipses. Although there is some indication that Z Camelopardalis may be an eclipsing system[3], no member of this small subgroup has, as yet, been investigated in the same manner as U Geminorum although we might argue that spectroscopically, all of the dwarf novae are very similar to U Geminorum and certainly there is nothing in the comprehensive observational data obtained by Zuckermann[1] which is contrary to the hypothesis that it is the secondary component which is responsible for the observed eruptions. The implication of the recent photoelectric work by Krzeminski[7] would seem to be that we may assume, for the purposes of the present discussion, that this is also the case for the Z Camelopardalis variables.

The Secondary Components of the Z Camelopardalis Variables

Suppose we first examine some of the physical characteristics of the secondary components of the Z Camelopardalis star, as far as we know them, and compare them with the secondaries in the U Geminorum systems. By this means we may be able to discover some parameter which may be the cause of the 'standstills'. Now it appears that the absorption spectrum of the cooler companion is visible only in those dwarf novae which have relatively long orbital periods; those with periods of less than about 6 hours showing no indication of the cooler star's spectrum. We must therefore restrict ourselves, initially at least, to a study of only a small handful of the dwarf novae, namely AE Aquarii, SS Cygni, EY Cygni, RU Pegasi (all U Geminorum variables) and Z Camelopardalis. Of these, AE Aquarii is far from being a typical member of the dwarf nova class as we have seen, the outbursts being comparatively minor, of a highly irregular nature, of short duration and possessing a very short mean cycle. In addition, the orbital period, although not known with the same degree of accuracy as for other members of the class, would appear to lie between 10 and 17 hours.

Sufficient information has now been amassed to show that the masses of the secondary components of all these systems, including Z Camelopardalis, lie between 0.7 and 1.2 solar masses with the cooler component of Z Camelopardalis occupying a mean position within this range. There is, therefore, no reason for supposing that the mass of the secondary in the Z Camelopardalis system differs appreciably from those in the other variables just mentioned. Similarly, the secondary stars in all of the system investigated appear to fill their lobes of the equipotential surfaces with the result that mass is ejected through the inner Lagrangian point into the region around the blue star. Indeed, it would seem that the secondary components of both the U Geminorum and Z Camelopardalis classes form quite a homogeneous group and so far, no evidence has been produced to suggest that the two types differ to any appreciable extent. It would appear, from the observational data at present available, that the only parameter which is significantly different for the two classes of variable to warrant detailed investigation is that the orbit of RX Andromedae (a typical member of the Z Camelopardalis group) is almost certainly elliptical, all of the U Geminorum stars so far examined having circular orbits.

The Surface Temperature of the Secondary Component

Evidently, any interpretation of the 'standstills' must be extremely tentative owing to the meagre amount of information available and it would be quite wrong to suggest that an early solution of the many problems associated with the 'standstills' will be forthcoming. Since Z Camelopardalis is possibly an eclipsing variable, like U Geminorum, and is also about a magnitude brighter when at minimum, it is possible that a similar investigation to that carried out by Krzeminski[7] will, in time, provide us with a more detailed understanding of them. We must remember that it is unlikely that many of the Z Camelopardalis variables will be found which are eclipsing systems. Only about a score of these stars are known altogether and of these, only about half a dozen are sufficiently bright at minimum to be accurately examined by present photoelectric techniques. The chances of more than one or two of these being suitable candidates for detailed investigation are therefore exceedingly small. It may well be, therefore, that Z Camelopardalis itself will be the only system capable of such an investigation by the same method as used by Krzeminski on U Geminorum.

We can, of course, make some theoretical suggestions concerning this particular variable. As a starting point, we shall assume that the outburst takes the same form as that found for U Geminorum, namely that the cooler component increases its radius by approximately fifty per cent during this phase while at the same time the amount of gas which is ejected through the inner Lagrangian point increases due to the expansion of this star. The surface temperature of the secondary component will also increase sharply *but not to the same extent as that found for the U Geminorum system.* The reason for this is quite straight-forward. Whereas the amplitude of a typical outburst for U Geminorum is close to six magnitudes, that for Z Camelopardalis is only about three magnitudes, indicating a brightness increase during maximum of only sixteen-fold instead of a hundred-fold for U Geminorum. Now if we assume a surface temperature for the secondary at minimum light of 5,500°K which is in agreement with its spectral type of dG5 and make the further assumption that the radius of the secondary is 1.5 times greater at maximum than at minimum brightness, we may calculate the approximate surface temperature of the secondary star during an outburst by means of the following equation, ignoring any bolometric corrections, the magnitude of which are not easy to assess.

$$\frac{R_M^2 T_M^4}{R_m^2 T_m^4} = 16 \qquad (1)$$

where R_M and R_m are the radii of the secondary component at maximum and minimum brightness respectively and T_M and T_m are the corresponding surface temperatures. From equation (1) we find that the surface temperature of the secondary star during an outburst is approximately 11,000°K as compared with the 15,000°K found by independent methods for U Geminorum.

A Pattern of Events for the 'Standstills'

Following a normal outburst of a Z Camelopardalis variable, with the star returning smoothly to minimum immediately after the eruption, the secondary contracts to its normal dimensions and the surface temperature declines to that of a dG5 star. Can we now visualise a sequence of events which will lead instead to a 'standstill', bearing in mind that such a period of comparative quiescence may last anything from a few days to almost eighteen months and that the point of onset of such a 'standstill' appears to be virtually constant for any particular variable (in the case of Z Camelopardalis about one third of the way down the descending branch of the light curve)? Clearly, since we have assumed that the secondary component contributes most of the visible light of the system during a typical outburst, with the remaining ten or fifteen per cent being contributed by the blue companion and the brighter part of the ring of gas, the inescapable conclusion seems to be that the surface temperature of the dG5 star remains, for an unpredictable period of up to eighteen months, at some value intermediate between the 5,500°K when at minimum and the extreme temperature at maximum. Such a condition further presupposes that there is a continuous outflow of material at something only a little below its peak value through the inner Lagrangian point with the star retaining its relatively small increase in dimensions. In other words, the secondary companion remains in its abnormal state, or something very close to it, throughout the entire duration of the 'standstill'.

A very fundamental question now arises; one which must be answered before we can go very much further in our understanding of the mechanism of the 'standstills'. What is the basic cause of the instability in the cooler component which brings about these irregular pulsations producing the relatively small-scale expansions of the star resulting in the semi-periodic loss of the higher atmospheric levels through the inner Lagrangian point? One further extremely relevant question we must ask is clearly: 'What effect does the apparent ellipticity of the orbit have on this instability?'

The Pulsational Instability of the Secondary Star

The mere fact that the dwarf novae are close binary systems in which there is a flow of gas from one component to the other does not, in itself, explain why the secondary component should exhibit these peculiar irregular pulsations. A very large number of eclipsing systems are known, including the Algol and β Lyrae variables and also the W Ursae Majoris stars which may possibly be the precursors of the dwarf novae and in none of these systems do we find the curious outbursts or 'standstills' which are so characteristic of the U Geminorum and Z Camelopardalis variables. Indeed, if the dwarf novae are a later stage in the evolution of the W Ursae Majoris stars in which both components are yellow dwarfs and where the overall dimensions of the system are very similar to those found in the dwarf novae, it seems surprising that such instability does not arise in the primary component during the stage where it loses mass to the secondary. This fact in itself would suggest that the presence of a white dwarf companion may, in some way, be necessary to initiate the unstable pulsations.

According to Kraft and Luyten[8], all of the indications are that the secondary components of the dwarf nova systems lie on the main sequence and the evolutionary scheme suggested by them to explain why the ratio of the luminosities of the two components is a function of the orbital period indicates that as the red star loses mass (either by injection through the inner Lagrangian point or during the eruptions) it moves down the main sequence, maintaining at all times an equilibrium between the luminosity and its mass. Being main sequence stars, we cannot call upon the type of pulsation we find in the RR Lyrae and Cepheid variables to explain the irregular outbursts since these pulsating stars have evolved well to the right of the main sequence and there is evidence that the pulsatory mechanism is brought about by the onset of helium burning in the centres of these stars, a situation which does not hold for the red components of the dwarf novae.

One other theory, which has been put forward by Schatzman[9] to explain the outbursts of the novae, may be considered here in the case of the dwarf novae. Over a quarter of a century ago, the problem of non-radial oscillations being induced in one component of a close binary system was considered by Cowling[10] who showed that this was possible in certain types of polytropic configurations provided there is a lack of synchronism between the axial rotation of the star and the period of revolution of the other. Such modes of non-radial oscillation may then come into resonance with the orbital period. This idea, as extended by Schatzman suggests that in the case of the novae, the blue component may experience such oscillations, resulting in a conical ejection of mass in two preferred directions. The induction time between outbursts is, of course, relatively long in both the novae and recurrent novae where such a mechanism may exist.

Suppose now that a similar state of affairs exists in the dwarf novae systems but in this case it is the red secondary which experiences these non-radial oscillations with the preferred direction being in the direction of motion of the star in its orbit. Now although we know the orbital period of the Z

Camelopardalis system (7 hours), we unfortunately know very little of the period of axial rotation of the secondary component. According to the model proposed by Krzeminski for U Geminorum, the centre of mass of the system lies just within the surface of the red star closest to the inner Lagrangian point but in the absence of any observational data we cannot determine the axial inclination of the secondary to the plane of the orbit and hence its axial rotation. However, from a systematic study of the Algol and β Lyrae eclipsing variables, several hundred of which are known, a large amount of observational evidence has accumulated to show that in those systems where there is a gas stream present, the orbital period and the axial rotations of the two components are not synchronised; the gas stream carrying off mass and angular momentum from the larger, more massive star with the result that the smaller component rotates too rapidly and the larger companion too slowly for synchronisation. We would therefore expect to find a very similar situation for the dwarf novae.

There is also a second method whereby instability may be induced in the secondary star. The spectral types of the secondaries lying between dG5 and dK0, these stars have low surface temperatures under what we may consider as normal conditions (when the variables are at minimum brightness) and this being the case we may assume that their atmospheres are in convective equilibrium. As Paczynski[11] has shown, under these conditions, a star which fills its lobe of the equipotential surface may exhibit an instability on the pulsational time scale. As we shall see in the following chapter, when this occurs the star will oscillate about a point of balance with each cycle lasting for only a few hours.

The above considerations now make possible an approach to a probable sequence of events leading to a 'standstill' of a typical Z Camelopardalis variable although it must be borne in mind that subsequent investigations may modify this picture drastically. Following a maximum produced by a non-radial pulsation of the red star in which an excess of mass is ejected through the inner Lagrangian point and the surface temperature of the secondary rises by approximately 5,500°K, the variable begins to fade towards minimum. So far, the picture is identical with that found in the U Geminorum variables. In the absence of any other complicating factor, the variable will continue to fade to minimum and we find a typical outburst for a Z Camelopardalis star. There is, however, a complicating factor which we must now take into account, namely the ellipticity of the orbit which we have assumed for these stars. This means that not only do the velocities of the two components vary at different points in the orbit but the distance between their centres, and particularly that between the red star and the quasi-permanent ring of gas around the blue component, also vary.

Depending upon the eccentricity of the orbit it is quite possible that at certain times, when the two components are at the point of their closest approach *and during the outburst when the dimensions of the red star are at their greatest*, the latter is actually in physical contact with the brighter portion of the ring. Such a situation will not arise in the case of the U Geminorum variables where the orbits have been shown to be essentially circular.

At this point, we must examine the time scales involved in the various processes taking place otherwise we may overlook completely several vital factors. The outbursts of the Z Camelopardalis stars are, in general, shorter than those of the related U Geminorum variables. Nevertheless, Z Camelopardalis may remain at maximum for four or five days before commencing to fade, a period which corresponds to between twelve and sixteen orbital revolutions. Clearly then, during a typical outburst, the red star may come into contact with the gaseous ring at least twelve times, yet it is only on relatively rare occasions that a 'standstill' results. Quite obviously, it is not possible to explain the onset of a 'standstill' by the geometry of the system during the time the variable is at maximum.

If, however, during the return of the secondary component to normal, there is a certain critical point during the contraction of the red star where both close approach to the blue star and physical contact with the ring temporarily halts the contraction process, a 'standstill' will result. During such a phase, the outflow of mass from the secondary will remain essentially constant with the surface temperature too remaining at a higher than normal level. Some confirmation for such a restraining process comes from the observed facts that for each

Z Camelopardalis variable so far investigated, the 'standstills' begin at an almost constant point on the descending branch of the light curve and the spectrum of Z Camelopardalis obtained by Lortet-Zuckermann[12] when at 'standstill' shows the remarkable feature of broad, symmetrical absorption bands with greatly extended wings about a central emission.

It is still too early to offer any opinion as to the nature of this critical phase during the contraction of the secondary star. This must clearly await more detailed observations than are at present available. We may, however, examine the probable effects on the rotating ring of gas and the blue star. During a comparatively long 'standstill' we may expect matter to be transferred into the ring at a continously higher rate than normal resulting in changes in both the density of this material and its dimensions. This suggests too that the surface temperature of the blue star will be higher than normal during a 'standstill' since the proportion of material falling into the atmospheric levels will be increased. This possibility has an interesting consequence. Whereas during a normal outburst, the blue star and the brighter portion of the ring contribute only between ten and fifteen per cent of the total light of the system, this proportion may be somewhat higher during a 'standstill', inferring that we do not require to postulate as high a surface temperature for the secondary component as would otherwise be the case.

The Irregular Light Behaviour of the Z Camelopardalis Stars

One peculiarity of the Z Camelopardalis variables still remains. We have still to explain those periods of completely erratic light variation which are also characteristic of these stars. During these periods they seldom attain the extremes of brightness shown during more normal behaviour. The explanation of these anomalous fluctuations in brightness may presumably be sought in highly irregular pulsations in the red secondary star. Whether these may be due, in part, to interaction of the outer atmospheric levels of this star with the rotating ring of gas is a question which cannot be answered at the present time. Doubtless further work on these variables will throw light on this important point.

REFERENCES

1 Zuckermann, M. C., *Ann. d'ap.*, **24**, 432 (1961)
2 Kraft, R. P., *Advances in Astronomy and Astrophysics*, **2**, 43, ed. Z. Kopal, (New York: Academic Press) (1963)
3 Kraft, R. P., *Ap. J.*, **135**, 408 (1962)
4 Holborn, F. M., *J. Brit. Astr. Assoc.*, **70**, No. 3, 134 (1960)
5 Glasby, J. S., *J. Brit. Astr. Assoc.*, **79**, No. 2, 146 (1969)
6 Glasby, J. S., *J. Brit. Astr. Assoc.*, **76**, No. 5, 351 (1966)
7 Krzeminski, W., *Ap. J.*, **142**, 1051 (1965)
8 Kraft, R. P., and Luyten, W. J., *ibid*, **142**, 1041 (1965)
9 Schatzman, E., *Ann. d'ap.*, **21**, 1 (1958)
10 Cowling, T. G., *M.N.*, **101**, 368 (1941)
11 Paczynski, B., *Acta Astr.* (1965)
12 Lortet-Zuckermann, *Ann. Astrophysics*, **29**, 205 (1966)

11. Gas Streams in Close Binary Systems

So far in the book we have dealt with the more general features of the dwarf novae and their galactic distribution, their relationship to various other classes of variable stars, the methods used for investigating their intrinsic physical characteristics and the results which have been obtained by the application of these new methods. It may be emphasised here that in spite of the vast amount of published information pertaining to these peculiar variables, this represents investigations made on only a small sample of the total number of dwarf novae known at the present time. Mainly, of course, this is due to their general faintness, even at maximum brightness. The task of gathering data for the production of extended light curves is very difficult and time-consuming for the majority of these stars and that of obtaining satisfactory spectra even more so.

For this reason we are often forced to extrapolate a good deal and it is generally gratifying to find that additional information, as it becomes available, agrees well with many of the predictions which have been made in this way. Following upon the spectroscopic work of Joy[1,2], Walker[3,4] and Kraft[5] there now seems little doubt that most, if not all, of the dwarf novae are close binary systems similar in some ways to the old novae, and in others to the W Ursae Majoris variables. The various arguments which have been put forward to support one or other of these relationships have been discussed in Chapters Seven and Eight. However, several important points were raised then which we shall now examine in detail.

As we have seen, one very important feature of the dwarf novae is the presence of gas streams in these systems, a feature they have in common with other classes of eclipsing binaries, notably the β Lyrae stars. There is sometimes a tendency to regard these moving gas streams as a third component of these binary systems, detached from the two principal components. This was done to a certain extent when we were discussing the outbursts of the dwarf novae. It would be wrong, however, to take this view too literally. The gas streams perform one function which is of great importance in the evolution of these binaries – that of transferring mass from one component to the other. As we shall see, this exchange of mass is closely associated with their evolutionary development and the inferences which may be drawn from the dynamics of such gas streams are very far-reaching, leading to many important consequences which we shall now discuss in detail.

Let us first consider one problem which arises from the hypothesis that the dwarf novae are generically associated with the W Ursae Majoris variables. As we have already seen, quite a large number of each class of star are known and of these, several have been extensively studied spectroscopically. Yet no variables have so far been discovered which possess properties intermediate between the two. Now this is clearly a point of fundamental importance since, as Kopal[6] has pointed out, this cannot be explained on the basis of observational selection. If, as has been suggested, the U Geminorum stars are a later stage in the evolution of the W Ursae Majoris variables, we would expect to find some stars (albeit only a small number), which are in the process of transition from one type to the other. The fact that none have been discovered would imply either that our original hypothesis is wrong, or that this intermediate stage is so short-lived that the chance of finding a variable in the process of transition is extremely remote. In order to understand the arguments which have been advanced to support the idea that evolution through the transition stage is unusually rapid, we must first discuss the different time scales in which these binary systems may evolve.

Evolutionary Time Scales

Our evolutionary picture of close binary systems has altered appreciably since the beginning of the century and in particular there has been the recognition of three different time scales which, as we shall see, are dependent upon the structural changes taking place inside the stars.

To begin with, let us examine a star on the main sequence

whose composition is homogeneous throughout. Such a star will consist mainly of hydrogen with perhaps seven per cent of helium, one per cent of other non-metals and less than half a per cent of metals, this representing the situation following the condensation of the star from the primeval gas cloud and before the nuclear reactions which begin during this early phase of its evolution have produced any marked change in the composition of the star. Once the nuclear processes begin in the centre of the star, where both pressure and temperature are greatest, hydrogen will be transmuted into helium, either by the proton-proton chain reaction or the carbon-nitrogen cycle, and a core of helium will be formed which brings about an increasing inhomogeneity of composition. The star will then commence to evolve towards the right of the main sequence and in doing so there will be an expansion of the outer levels of the star due to this change in internal constitution. The time taken for the star to evolve a significant distance from the main sequence is known as the *nuclear time* and is quite a slow process, being dependent upon the mass and luminosity. Thus, for a star of five solar masses it is of the order of 3×10^8 years.

If, however, the luminosity of the star is not in equilibrium with the energy produced, then the star will either expand or contract until a balance is achieved. For example, if the amount of energy generated within the star is too high for its luminosity, the star will expand while at the same time the internal temperature will fall, the converse occurring if the generation of energy is too low to balance the luminosity. This self-governing period is known as the *Kelvin time scale* of the star and is about 200 times shorter than the corresponding nuclear evolution period. For a star of five solar masses, the Kelvin time is about 1.25×10^6 years.

Just to complete this discussion, we must consider a third time scale, known as the *pulsation time* which applies to those stars where the gravitational force is not balanced by the pressure in the central regions of the star (the latter being due mainly to the high temperatures present in this region). If this should happen, we find that the star will oscillate about a new point of balance, each cycle lasting for only a few hours. Compared with the other two evolutionary time periods, the pulsation time scale of a star is therefore very short indeed.

In the present chapter, we shall be concerning ourselves only with the nuclear and Kelvin time scales and it will readily be seen that if, during the normal evolution on the nuclear time scale, a transition process occurs which is both unstable and governed by a Kelvin time scale, this will satisfactorily explain why no intermediate systems have so far been discovered since evolution through a Kelvin time period is, as we have seen, at least 200 times more rapid than on a nuclear time scale. As a result, we would expect to find, at most, only one system in more than 200 which is at present in the process of transition. As only about a hundred of these particular close binaries have been studied in detail, it is therefore not really surprising that no intermediate types have been found.

Analyses of several binary systems, mostly of the Algol and β Lyrae types, have been made by Kopal[1,7,8] and Crawford[9] and these have revealed one distinguishing feature which ties in well with what has just been said. Among all of these binaries, there are none in which the more massive primary component fills its lobe of the equipotential surface. It is always the secondary which has expanded to fill the corresponding lobe. Formerly, this would have indicated a very peculiar course of evolution for we must almost certainly conclude that the components of these close binaries are of approximately the same age, having been born out of the same gas cloud (the chance of two individual stars coming together and forming such systems is exceedingly remote) and on this basis, the more massive of the two would be the first to begin evolving to the right of the main sequence, expanding in the process. But we now see that if we can postulate some means by which the initial exchange of mass occurs on a Kelvin time scale, then both of the problems we have just considered disappear.

An investigation into this question of mass exchange during the evolution of close binaries has recently been made by Morton[10] with very interesting results. It will be recalled that in the Introduction, mention was made of the equipotential surfaces of a close binary system which, in a two-dimensional diagram, project into the familiar figure of eight (Fig. 1). Now unless the two stars are very widely separated, each component will, in turn, fill its lobe of the equipotential surface on the long, nuclear time scale during the normal course of its evolution. The more massive primary star will, according to the accepted theories of stellar evolution, be the first to begin evolving away from the main sequence and it is generally thought that this component will fill its lobe of the equipotential surface before any marked changes have taken place in the secondary companion. This will certainly be true in the case of stars of unequal mass (we have already seen in Chapter Seven that the radial velocity curves of the W Ursae Majoris variables indicate that the primary is about three times more massive than the secondary), and once the primary has filled its lobe, further expansion will result in a flow of gas through the inner Lagrangian point towards the secondary. During this phase of evolution, therefore, the secondary is collecting the mass which is being lost by the primary with the result that eventually, the former becomes the more massive of the two. The situation is now reversed and it is the secondary which fills its lobe while the primary evolves back towards the main sequence.

Morton[10] has suggested that once mass begins flowing from the primary through the inner Lagrangian point and into the lobe of the secondary, an unstable situation arises which removes the evolutionary process from the nuclear time scale and places it on a Kelvin scale. If this is so, then the evolutionary phase in which mass transfer takes place will be quite rapid. At this point, taking into account the importance of this instability in placing the evolutionary process on a Kelvin time scale, we must discuss it in a little more detail. Why, for example, should mass loss from the primary be associated with instability in these binary systems?

This particular question has been investigated by Morton[10] who has calculated the effect which the removal of mass from the outer atmospheric levels of various model stars will have on their radii and, equally important, what happens to the primary lobe of the equipotential surface as mass is being transferred. For a considerable number of these binaries, we find that before the primary has evolved to the point where it completely fills the lobe, it enters the region on the Russell-Hertzsprung diagram known as the Hertzsprung gap. Very few stars, even normally evolving single stars, are found here since evolutionary expansion is quite rapid in this region. It follows then that most of the close binaries undergo mass exchange within the Hertzsprung gap and consequently the chance of finding a system in which the primary fills its lobe of the equipotential surface is very small indeed.

Morton[10] has also considered the case where the primary star has expanded to fill its lobe before reaching the Hertzsprung gap. When this happens we cannot, of course, apply the argument given above to explain the deficiency of intermediate systems and we must therefore look for some other explanation. This is to be found in the instability mentioned earlier, an instability which arises once mass exchange begins and after thermal equilibrium has been attained by the primary.

At this point, we must introduce two different models for such a star. In the first there is a discontinuity in the physical structure, the star consisting of a homogeneous core which is rich in helium, an atmosphere which is similarly homogeneous but hydrogen-rich, and an intermediate zone which possesses the same composition as the core but in which energy is transferred by radiation rather than by convection which is the case for the core. In the second type, the composition is continuous. As before, we have a convective helium-rich core and a radiative hydrogen-rich atmosphere but the intermediate zone, although again radiative in nature, possesses a composition in which the amount of hydrogen increases steadily as we move away from the centre.

In these models where the composition is discontinuous, Morton has shown that the radius will increase as mass is lost through the inner Lagrangian point, whereas it decreases for stars with composition continuity. In both instances, however, the predominant process is not the change in the radius of the star but a marked shrinkage of the dimensions of the lobe of the equipotential surface, this being responsible for the instability on a Kelvin time scale.

Here it may be asked: How long does this unstable period last? Does it continue throughout the entire period of mass exchange, or only during the early stages of this phase of evolution? The answer is that, theoretically, it may be shown that a return to stability and to a nuclear time scale will occur only when the primary has lost sufficient mass for it to return to a homogeneous condition, although now with a higher helium content than before since a high proportion of hydrogen will have been lost from the outer layers. This condition will only be attained after the secondary has filled its lobe and the roles of primary and secondary have been reversed.

REFERENCES

1 Joy, A. H., *Publ. A.S.P.*, **52**, 324 (1940)
2 Joy, A. H., *Ap. J.*, **124**, 317 (1956)
3 Walker, M. F., *Publ. A.S.P.*, **66**, 230 (1954)
4 Walker, M. F., *Ap. J.*, **123**, 68 (1956)
5 Kraft, R. P., *ibid*, **135**, 408 (1962)
6 Kopal, Z., *Ann. d'ap.*, **19**, 298 (1956)
7 Kopal, Z., *ibid*, **18**, 379 (1955)
8 Kopal, Z., *Close Binary Stars* (New York: John Wiley & Sons, Inc.) (1959)
9 Crawford, J. A., *Ap. J.*, **121**, 71 (1955)
10 Morton, D. C., *ibid*, **132**, 146 (1960)

PART TWO

12. U Geminorum Variables

In the first part of this book we have dealt with the general characteristics of the dwarf novae where we were primarily concerned with a discussion of this class as a whole; the discovery of these peculiar variables, their distribution within the galaxy and in particular with their light and spectral changes which finally led us to the conclusion that they are binary systems very like the old novae and the W Ursae Majoris stars. We have also seen how their close similarity to the latter class of variable has led many astronomers to believe that the dwarf novae are a later stage in the evolution of the W Ursae Majoris stars.

Typical light curves have been given in the text for a small number of the dwarf novae but these have been merely to illustrate some major peculiarity in a particular variable. In this second part therefore, we shall be examining the known members of this fairly small class individually, although it must be appreciated at the outset that, owing to the general faintness of these objects, there are many for which very little is known with certainty. Even such a basic property as the light curve is known for only a small proportion of them and much more work is required, using the largest instruments available, before satisfactory spectra are obtained. It is to be sincerely hoped, however, that the data given in this, and the following chapter, will stimulate further research on these interesting and very important stars.

In the present chapter we shall consider those variables which have been assigned to the U Geminorum class of the dwarf novae; that is, those for which there is, at present, no evidence of any 'standstill' having occurred, the star fading fairly smoothly to a normal minimum after each outburst. The much smaller subgroup of the Z Camelopardalis variables will be dealt with in detail in the next chapter. At this point we must, of course, bear in mind that as more data accumulates there is the possibility that some of those stars now recognised as U Geminorum variables (especially the fainter members and those which have been only sporadically observed) may have to be transferred to the Z Camelopardalis subgroup. As we shall see, some of the stars of the latter group have fairly frequent periods of 'standstill' while there are others in which they occur only rarely and consequently long periods of observation are required before we can be reasonably certain that no 'standstills' do happen. Some of the assignments given below, therefore, may prove to be invalid as more information becomes available.

The values given for the maxima and minima of the various stars are the extremes which have been recorded and we should not presume that the variable will attain either of these extremes at every outburst for although the minima are, in general, fairly constant, wide variations are common for the maxima. It is usually found, for example, that the typical long maxima are brighter than the shorter variety, often by as much as a whole magnitude. In many instances, too, the star is so faint at minimum that its magnitude at this phase has not yet been recorded. Most of the magnitudes given in this chapter are photographic as the majority of the U Geminorum variables have been discovered on patrol plates or during photographic surveys of selected Milky Way fields; in those cases where the visual magnitudes have been well established by extended series of observations, these are given in preference to the photographic values.

Another difficulty encountered in the study of these variables is that of obtaining satisfactory spectra of the fainter members of the class. This is much more time-consuming than the estimation of visual or photographic magnitudes and our knowledge of the spectroscopic features of the dwarf novae is confined to only a handful of spectra taken of the brighter variables. Indeed, it is a triumph of modern astronomical technology that high-dispersion spectrograms have been obtained of stars as faint as 15.4 magnitude.

In order to facilitate reference to the variables in this chapter, they have been arranged in alphabetical order of the constellations, this being considered simpler than an arrangement in order of right ascension.

AR ANDROMEDAE

$01^h\ 42^m\ 04^s$ $+37°\ 41'.7$ (1950.0) $11^m.5$–$16^m.3$ (photographic)

For many years, this variable was thought to be a typical long period variable with a mean period of 264.5 days although no details of its spectroscopic characteristics, and few observations of its light variation, have been recorded. However, Makarian[1] has recently re-examined this star and shown that both the light changes and the spectrum are consistent with it being a U Geminorum variable with a mean period of about 68 days. There is a rather strong ultra-violet continuum which is present throughout the whole of the light cycle and the emission lines, unlike those of the long period variables, are observed only after a noticeable decrease in brightness and during the ensuing minimum. In a typical long period variable, the emission lines appear at some point during the rise to maximum and generally fade quite swiftly once the light begins to decline and are rarely, if ever, found at minimum, the spectrum at this phase being dominated by broad absorption lines of either titanium or zirconium oxide.

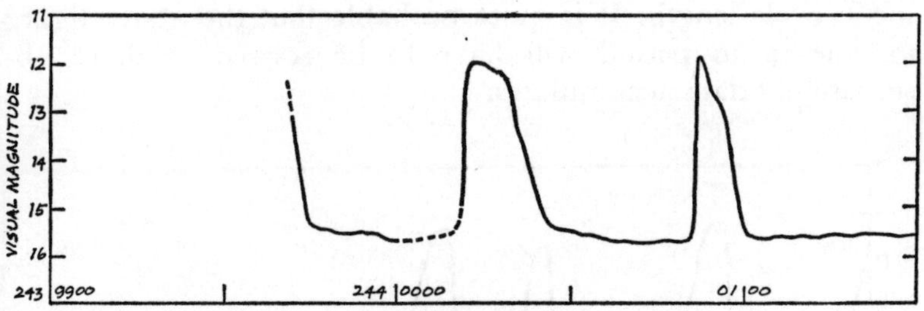

Fig. 21 – Light curve of AR Andromedae.

The light curve (Fig. 21), drawn from 183 observations by the author made during 1968, confirms the shorter period given by Makarian and also indicates the abrupt rises of this star which are so characteristic of a U Geminorum variable. Since this star attains a visual magnitude brighter than 12.0 at maximum, it is well within the reach of moderate-sized instruments, although large apertures are necessary to follow the variable through its minima. Fig. 21, although of a very preliminary nature, and based upon a relatively small number of observations made over a fairly short period, does show the existence of two types of maxima, long and short, similar to those of U Geminorum itself.

VZ AQUARII

$21^h\ 27^m\ 48^s\ -03°\ 12'.6$ (1950.0) $11^m.8$–$15^m.0$ (photographic)

Although this variable reaches a visual magnitude at a normal maximum of 12.3, it appears to have been comparatively little observed and few details of its light curve are given in the literature. The light curve given in Fig. 22 is a portion of that derived by the author from 577 observations made during the period 1965–68 and indicates a mean period of approximately 40 days. The field of the variable is readily found from the 5.69 magnitude star 21 Aquarii but being a zodiacal variable it is unobservable throughout three of the winter months and this, taken in conjunction with its low southerly aspect during the summer months in European latitudes, means that it is not easy to observe at sufficient successive maxima to provide a really reliable estimate of its mean cycle length. It is quite probable that the above figure for the mean period will have to be revised as more observational data accumulates.

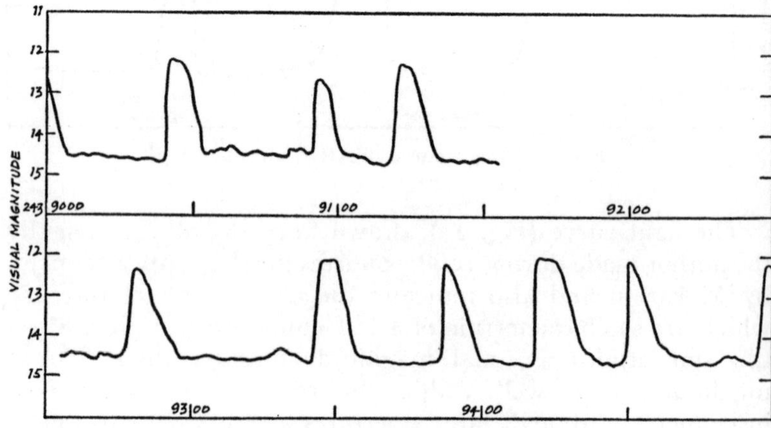

Fig. 22 – Light curve of VZ Aquarii.

The rises of this variable are, in general, quite steep, the decline to minimum often showing pronounced irregularities. Fig. 22 also shows certain other characteristics of VZ Aquarii, notably the wide deviation from the mean period over individual cycles and the variation in the maximum brightness attained at different maxima. The minima, too, are sometimes highly disturbed with fluctuations of the order of 0.4 magnitude being commonplace. As with the majority of these variables, the long, flat maxima tend to be brighter than the shorter ones.

AE AQUARII

$20^h\ 27^m\ 33^s$ $-01°\ 03'.0$ (1950.0) $9^m.7$–$11^m.7$ (photographic)

It is now generally accepted that AE Aquarii is a member of the U Geminorum class but quite clearly, from its light curve (Fig. 23), it is far from being a typical representative.

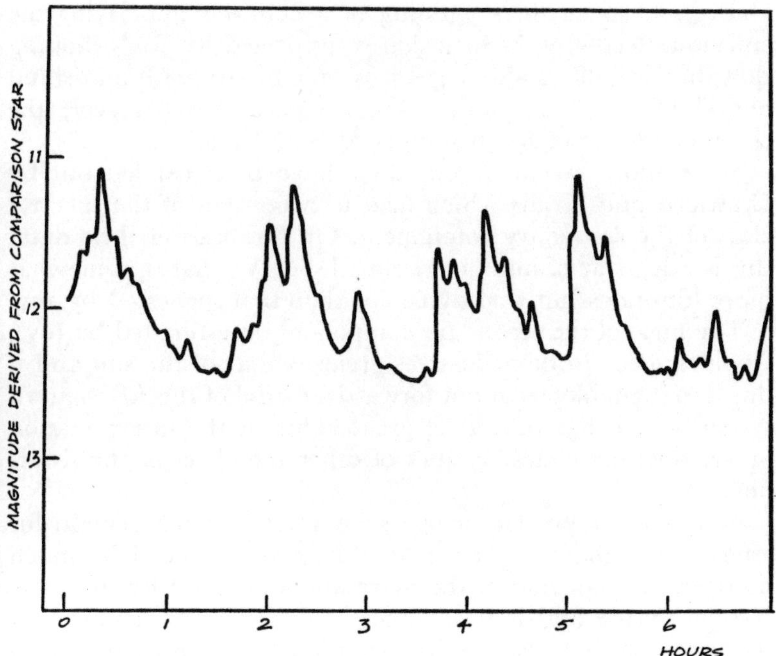

Fig. 23 – Photoelectric light curve of AE Aquarii.

The more usual light fluctuations are extremely rapid and complex and are almost invariably less than half a magnitude in amplitude; a situation which has resulted in the production of photoelectric (rather than visual) light curves for this star.

More violent outbursts, with an amplitude of up to about two magnitudes have, however, been observed by Zinner[2] who suggested its inclusion in this class of variable. Although he has recorded three such outbursts with a period of about 371 days superimposed upon the much smaller, apparently semi-regular fluctuations, it is quite clear from observations made over several decades that maxima of this amplitude must be extremely rare. On what evidence, therefore, do we base its inclusion among the dwarf novae? The answer to this is two-fold. Firstly, the spectrum of AE Aquarii when at minimum bears a very strong resemblance to such stars as RU Pegasi; consisting of strong emission lines of hydrogen and singly ionised calcium about 20 Å in width with a somewhat fainter emission due to neutral helium. There is also a very prominent absorption spectrum suggesting of a dG8 star underlying the emission. Secondly, its inclusion is supported by Joy's[3] finding that the variable is also a spectroscopic binary with an orbital period of about 17 hours. According to this observer, the components are of spectral types sdBe and Ko.

Some more recent observations have been carried out by Crawford and Kraft[4] which lead to a revision of the spectral class of the secondary component. On the basis of their data, this is a K5 star of luminosity class IV or V; that is, somewhat more luminous but slightly cooler than that suggested by Joy.

The mass of the secondary component, as estimated by Joy[3], would appear to be at least as great as that of the sun and if this is so it enables us to put forward a model of the AE Aquarii system which has proved of great value in the interpretation of the physical characteristics of other members of the dwarf nova class.

Now a K5-type star of one solar mass will not remain for long on the main sequence since its evolution will be much more rapid than that of the more abundant, but less massive, K-type dwarfs and it seems reasonable for us to assume that this star has now begun its evolution away from the main sequence and towards the right of the Russell-Hertzsprung diagram. During this process, the size of the helium core in the centre of the star will increase and the star itself will expand. Indeed, there is now quite a lot of observational data to show that this component has almost certainly filled its lobe of the equipotential surface and is now ejecting mass through the inner Lagrangian point; matter which, in the form of a stream of gas, has entered the corresponding lobe of the blue star and has formed a ring, or disc, around the primary.

The spectrum of AE Aquarii clearly shows the presence of sharp emission lines of singly ionised calcium which have the same radial velocity as that of the blue star and are obviously closely associated with it in its orbit. In addition to this, we find that the blue companion itself shows certain curious anomalies. For example, as we have mentioned earlier, if we plot its position on the Russell-Hertzsprung diagram, we discover that it occupies a place at least ten magnitudes below the main sequence among the white dwarfs and yet it is slightly more than four magnitudes more luminous than a typical white dwarf star. Clearly there is some physical process taking place within the system which is making an important contribution to the surface brightness of this star.

The most plausible explanation of this has already been briefly discussed in Chapter Two but, since it leads us to a possible cause of the minor outbursts of this variable, it will be examined here in more detail. As a first step, let us examine the temperature and density of the gas stream which is present within the lobe of the white dwarf component. The temperature is more easily discussed since it will not vary appreciably from the surface temperature of the cool companion from which it originates. Now the average surface temperature of a K5 star is about 4,000°K and we must, of course allow for some initial cooling of the gas due to its expansion as it passes through the inner Lagrangian point. Within the ring itself, however, there will be a heating process since the gas will absorb ultra-violet quanta from the blue star, this then being emitted as a longer wavelength by the process of fluorescence, a certain proportion of the visible light of this gaseous ring being due to fluorescence. It is therefore reasonable to assume that the temperature of the rotating ring around the blue component is between 4,000°K and 5,000°K.

It is more difficult to estimate the density of the gas stream. Due to the injection of mass into this ring (a process which may not even be continuous), the probability is that the density is not constant throughout the ring. We would not expect the average density to be much greater than that of the outer levels of the K5 star and quite possibly it is much lower than this.

Now in Chapter Eleven, we saw how, during the evolutionary career of a close binary system, there is an interchange of mass from the primary to the secondary, this mass exchange taking place on a Kelvin time scale in a relatively short period when compared with the much longer nuclear time scale. Eventually, the direction of mass flow will be reversed as the primary evolves back towards the main sequence and then down towards the bottom left-hand corner of the Russell-Hertzsprung diagram, possibly passing through a red giant stage and then collapsing swiftly to a white dwarf, while the secondary component expands to fill its lobe of the equipotential surface. This is the situation we generally find in the case of AE Aquarii and many other of the dwarf novae.

It is therefore quite possible that heating of the white dwarf star occurs due to the accretion of mass from the rotating gas stream within its lobe. Now the rate of accretion is given by the following expression assuming that all of the kinetic energy of the infalling material is converted into heat,

$$\sigma = \frac{RL}{G\mathfrak{M}} \tag{1}$$

where R, L and $\mathfrak{M}$ are respectively the radius, luminosity and mass of the white dwarf and G is the gravitational constant. We must, of course, make certain assumptions here about the mass of the blue primary but if we accept a figure of 0.7 solar masses which is in good agreement with an orbital period of approximately 17 hours, the value of σ turns out to be of the order of 1.0×10^{22} kilograms annually.

Up to this point we have made the assumption that the secondary is ejecting mass through the inner Lagrangian point merely by virtue of its evolutionary expansion. We must now consider how far we are justified in taking this view. Are there, for example, any other mechanisms by which mass may be lost through the inner Lagrangian point?

Much will naturally depend upon the mass of the secondary, whether it is sufficiently great for appreciable evolution away from the main sequence to have taken place. In the case of most dwarf novae, this is almost certainly not so, for these systems have secondaries whose absolute magnitudes are between +8.0 and +10.0. It may be, however, that AE Aquarii is an exception since the secondary in this system has an absolute magnitude of +7.0 and, as Crawford and Kraft[4] have pointed out, such a star will eject mass at this rate, or even higher, due to evolutionary expansion. The above argument depends entirely upon the deduced mass of the cool companion being large enough for helium formation in the core to have reached a fairly advanced stage and recent estimates of the orbital period of AE Aquarii by Walker[5] suggest that it is much shorter than the earlier accepted value of about 17 hours, possibly being as short as 10 hours. This would place the mass of the secondary in the same range as those of the other dwarf novae, namely about 0.7 solar masses. As a result we may have to turn to one of the alternative modes of mass ejection described in Chapter Four; non-synchronisation of orbital and axial rotation, emission of gravitational waves, stellar winds or irregular pulsation in the outer layers of the red star. The last mentioned method is possibly operative to a certain extent in this variable since it has, as we have seen, been shown to be the cause of the major outbursts in U Geminorum and although we have no direct observational proof of it in this particular system, there is no evidence against it. On this interpretation, the pulsations present in the cool star will, of course, be less pronounced than those found for U Geminorum.

CV AQUARII

$21^{h}\ 18^{m}\ 58^{s}$ $-14°\ 31'.0$ (1950.0) $12^{m}.4$–$16^{m}.9$ (photographic)

CV Aquarii was discovered by Shapley and Hughes-Boyce[6] on plates taken in 1933–34 during a photographic survey of this constellation. It is readily located from the 5.97 star 16 Aquarii but being extremely faint when at minimum, this phase has been relatively little observed. The light curve given in Fig. 24, is based upon 310 visual observations made by the author during the period 1967–68 using a 13-inch reflector.

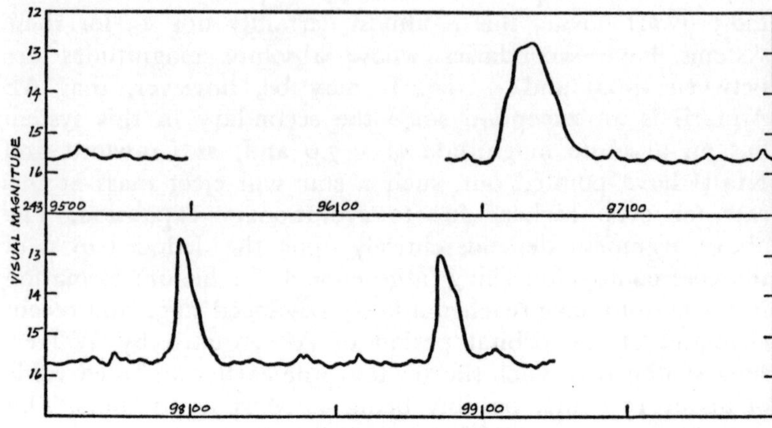

Fig. 24 – Light curve of CV Aquarii.

Since this variable lies close to the ecliptic, it is unobservable between October and early January when the sun is passing through this zodiacal constellation and many maxima are unavoidably missed. The amount of published data on this star is, at the moment, not sufficiently comprehensive for a mean period to be assigned to it and as yet we know very little of its spectrum. With a photographic magnitude at minimum as faint as 16.9, it will obviously be a difficult matter to obtain satisfactory spectra at this phase, even with the 200-inch reflector at Mount Palomar. At maximum, the star is well within the range of moderate-sized instruments.

The star field is sparsely populated unlike those of the U Geminorum variables which are situated in, or close to, the Milky Way and when at maximum there is little difficulty in identifying the variable.

UU AQUILAE

$19^h\ 54^m\ 35^s$ $-09^\circ\ 27'.4$ (1950.0) $11^m.4$–$15^m.9$ (visual)

UU Aquilae is a typical U Geminorum variable which has been extensively observed for many years by the American Association of Variable Star Observers[7] and more recently by the Variable Star Section of the British Astronomical Association. The minima are quite faint but there is some evidence that the brightness of UU Aquilae at minimum varies to a greater degree than in most of the other dwarf novae which have been investigated. Sometimes, the star does not fade much below 14.7 magnitude while other instances have been recorded where it has been as faint as 15.9 magnitude although the latter are apparently more common occurrences than the former. The photographic range has been given as 11.0 to 16.8 magnitude.

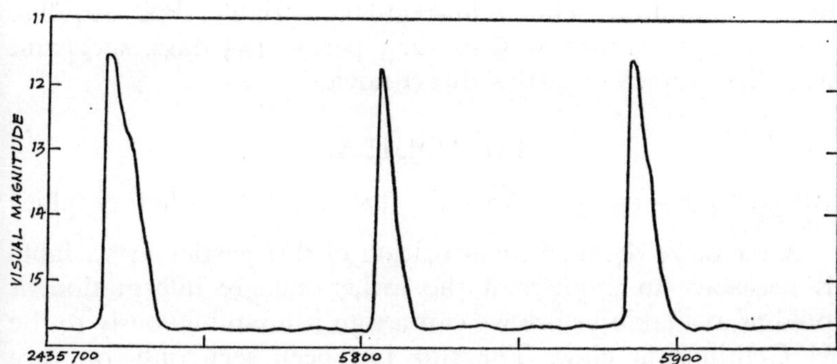

Fig. 25 – Light curve of UU Aquilae (*Courtesy of the British Astronomical Association*).

The light curve of this star is given in Fig. 25 and again it will be seen that both long and short maxima occur. Although there is a general tendency for these to alternate, a feature which is also characteristic of U Geminorum, this is not an invariable rule and two or more maxima of the same type have been known to follow each other. Even in the case of the long maxima, however, the time spent at maximum is fairly short and there is often a distinct hump on the descending branch of the curve. Deviations from the mean cycle of 56.2 days are normally quite small but an interval as short as 31 days has been recorded between two maxima.

In the course of their comprehensive survey of the spectra of dwarf novae and related variables, Elvey and Babcock[8] examined this star when at minimum employing a low dispersion of approximately 340 Å per millimetre and fairly long exposures. At this phase, the spectrum shows a moderately strong emission with Hγ being the most dominant line. In addition to this, there is also an underlying continuum whose intensity distribution is similar to that of a G-type star. No indications of an absorption spectrum were found in the photographic region.

The field of UU Aquilae may be conveniently found from the bright pair of stars, 56 and 57 Aquilae, the latter being a close optical double. Several faint stars of between fourteenth and fifteenth magnitude lie close to the variable and make identification difficult whenever the star is passing through a deep minimum. The semi-regular variable PX Aquilae (photographic range 10.9 to 12.4, period 155 days, spectrum M5) lies very close to this dwarf nova.

FH AQUILAE

$18^h\ 59^m\ 43^s\ -05°\ 41'.1$ (1950.0) $13^m.4$–$(16^m.0$ (photographic)

A far more detailed investigation of this particular variable is necessary to implement the rather meagre information at present available before we can assign it unambiguously to the U Geminorum class. The star has been seen only on rare occasions and then only when around maximum brightness and consequently there is still some doubt as to its real nature. We cannot even rule out the possibility that it may even be a recurrent nova since at the present time there is insufficient data to provide even an approximate figure for its mean period. As the magnitude at minimum is not accurately known, the amplitude too cannot be determined with any degree of accuracy.

The suggestion that it may be a member of the U Geminorum class was put forward by Himpel[9] in 1944. From the scanty details which we do have of its light variation, the rises would appear to be both infrequent and of the abrupt type characteristic of the dwarf novae (or for that matter the recurrent novae). Owing to its general faintness, large apertures are necessary for the observation of this variable, even when at maximum. If it is a true U Geminorum variable, it would appear to most strongly resemble SW Ursae Majoris, both in amplitude and period.

The star lies within the boundaries of the Milky Way, on the border with Scutum, the field being fairly readily found from the red, fourth magnitude star 12 Aquilae. There are, unfortunately, several faint stars close to the variable of about the same magnitude as FH Aquilae when at maximum, making positive identification very difficult. No spectroscopic details are known.

FO AQUILAE

$19^h\ 14^m\ 04^s$ $-00°\ 02'.3$ (1950.0) $13^m.5$–$16^m.1$ (photographic)

This dwarf nova has been studied by several observers, in particular by Morganroth[10], and in general it exhibits typical long and short maxima very similar to those of the type star, U Geminorum. None of the anomalous maxima, characteristic of SS Cygni, have been recorded with certainty but we must remember that this faint member of the class has been relatively

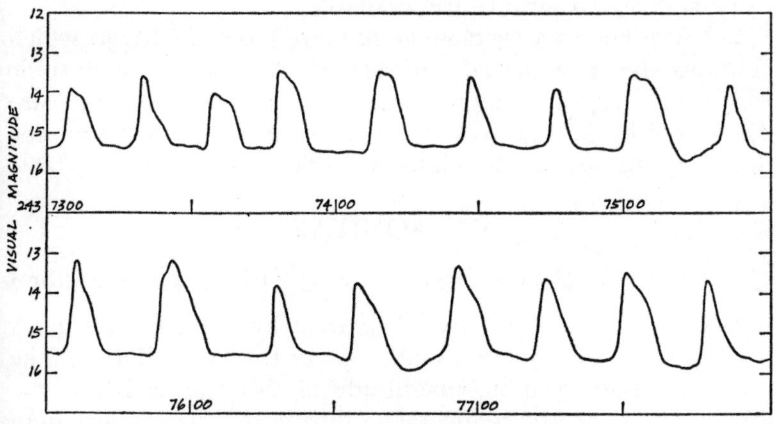

Fig. 26 – Light curve of FO Aquilae.

little observed since its discovery and although the light curve (Fig. 26) is fairly complete, this covers only a short period of approximately two years and the existence of such anomalous maxima cannot be entirely ruled out.

The mean cycle of FO Aquilae is quite short for a U Geminorum variable, that suggested by Morganroth[10] of only

28.7 days, being substantiated by the observations made by the author (Fig. 26). The visual range would appear to be 13.1 to 15.7 magnitude, the minima being, in general, quite smooth (although at the very limit of the instrument used) but at least two minima, about 0.3 magnitude deeper than normal, were recorded. As will be seen from the light curve, the maxima too, vary quite considerably in brightness and with this star, there is not the same tendency for the long maxima to be substantially brighter than the short variety; indeed, the first two long maxima recorded were appreciably fainter than the short, sharply-peaked kind. The presence of a distinct hump on the descending branch of the light curve, a fairly common feature of this class of variable, will also be noted, this rarely being seen on the rise from minimum.

So far, very little has been published of the spectroscopic features of this star. At minimum, of course, it is possibly beyond the present limits of detection and we must await further refinements of spectroscopic techniques before high-dispersion spectrograms are available.

FO Aquilae lies very close to Barnard's nebula B 138 which contains the long period variable FP Aquilae (photographic range 11.1 to 16.6 magnitude, mean period 338 days) also discovered by Morganroth. Being another Milky Way star, the field is quite densely populated with faint stars.

KX AQUILAE

$19^h\ 31^m\ 36^s$ $+14°\ 11'.0$ (1950.0) $12^m.5$–($16^m.3$ (photographic)

KX Aquilae was discovered in 1932 by Hoffleit[11] during a photographic survey of variable stars in this constellation. The minimum photographic magnitude of this star is below the limit reached on the plates taken during this survey although from the light curve (Fig. 27), the visual minimum magnitude would appear to be about 15.6. The mean cycle derived from the observations is about 53 days but it must be emphasised that the period covered is extremely short and such a figure must be regarded as only approximate. Future data will almost certainly result in a revision of this value. As will be seen, the interval between successive outbursts varies quite appreciably.

Once again, we find the rise to maximum is more rapid and

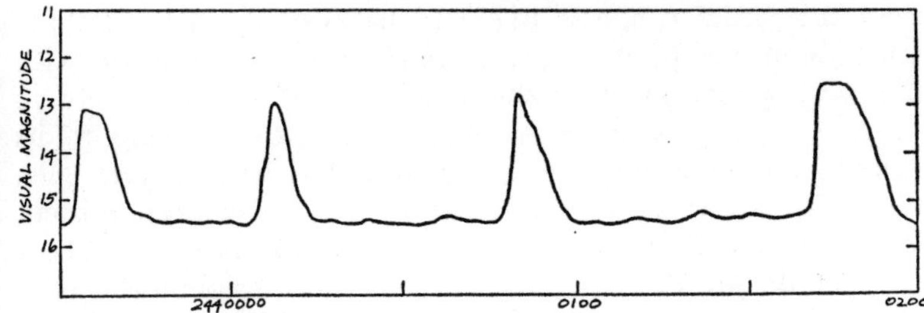

Fig. 27 – Light curve of KX Aquilae.

smoother than the decline and both long and short maxima have been recorded.

The variable lies in the main stream of the Milky Way where it crosses this constellation but the field is not as crowded as that of the previous star. When at maximum, this variable is a magnitude brighter than FO Aquilae, but again we know very little of its spectrum even at this phase of the light cycle.

V725 AQUILAE

$19^h\ 54^m\ 20^s$ $+10°\ 41'.2$ (1950.0) $13^m.7$–$16^m.2$ (photographic)

Found by Rolfe[12] in 1949, this variable does not appear to have been seriously studied since its discovery and no details of the light curve have been published. The mean period, too, is not known but if the star obeys the cycle-amplitude relationship, we would anticipate this to be in the region of 20 days, the photographic amplitude of 2.5 magnitudes being at the lower end of the range for the U Geminorum stars. Being so faint, the spectrum has not yet been investigated.

AT ARAE

$17^h\ 26^m\ 51^s$ $-46°\ 03'.8$ (1950.0) $13^m.0$–$15^m.8$ (photographic)

This particular U Geminorum variable was discovered by Shapley and Swope[13] in 1934 and belongs to the small group of these stars which have amplitudes apparently smaller than expected from the empirical cycle-amplitude relationship. The spectrum has not yet been investigated but if we assume that it is similar to those of the other U Geminorum stars, then the visual range is likely to be even smaller than the 2.8 magnitudes photographic, which is not compatible with the mean period given in the literature[13] of approximately 70 days. This, of course, may simply mean that, owing to its general faintness and the fact that it has not, as yet, been diligently observed, several maxima may have been missed. We have already seen that many of these stars exhibit short maxima during which the stay at maximum brightness is only of the order of one or two days (maxima which may easily be missed altogether due to bad weather and other causes, particularly when a photographic survey is carried out from a single observatory) and if this is so, then the mean period is almost certainly shorter than that accepted at present. The time during which such an investigation is carried out, moreover, is necessarily very short compared with that necessary for an accurate mean cycle to be derived for, as will also have become apparent now, the interval between successive outbursts can vary widely from the mean.

There is also one other possibility which we must not overlook. Although in the case of this particular star we have no direct spectroscopic proof (which will ultimately come from a determination of the orbital period), it is quite probable that the blue star is relatively fainter than the red companion, as is the case for RU Pegasi. The true visual amplitude may therefore be greater than the 2.8 magnitudes found observationally.

The position of the variable is not far from the 4.63 magnitude star σ Arae, within the southern Milky Way, is a very rich field of faint stars. Here again, we are faced with a difficult problem of positive identification when the star is around minimum brightness.

FV ARAE

$17^h\ 28^m\ 39^s$ $-63°\ 00'.9$ (1950.0) $13^m.0$–$(18^m.0$ (photographic)

Like many of the fainter southern U Geminorum variables, FV Arae was discovered photographically during surveys made of selected star fields in, or close to, the Milky Way. It must not be thought, however, that these Milky Way fields were examined expressly for the purpose of discovering dwarf novae.

As we have seen earlier, unlike the true novae, these variables show little tendency to concentrate along the galactic plane. The Milky Way fields, being exceptionally rich in stars, were chosen as the most likely to yield a large number of new variable stars of all types, a small number of which eventually turned out to be dwarf novae.

This particular variable was discovered by Swope[14] in 1935 and is certainly one of the faintest known U Geminorum stars at minimum. If we accept the mean figure of +7.5 for the absolute magnitude of this class of variable at minimum, this would seem to indicate that FV Area is one of the more distant members of this class.

Like the two preceding variables, FV Arae lies in a densely populated field and very few of its characteristics are known apart from its magnitude around maximum and the fact that the rise from minimum appears to be of the abrupt type normally associated with the dwarf novae. The photographic amplitude is quite large, in excess of five magnitudes and in the absence of any figure for the mean period, we might even be justified in considering this star as a recurrent nova, rather than a U Geminorum variable. Although the number of observations made is very small, the observational data would tend to indicate that the star is a true dwarf nova, possibly one with a fairly long mean cycle.

SS AURIGAE

$06^h\ 09^m\ 35^s$ $+47°\ 45'.5$ (1950.0) $10^m.5$–$14^m.5$ (visual)

SS Aurigae has been known for more than half a century, having been discovered in 1907 by Hartwig[15]. Being some two magnitudes fainter than SS Cygni, it has not been as fully observed as this star, but few maxima have been missed, especially during the past four decades. Full details of the light variations of this variable have been given in the Quarterly Reports of the American Association of Variable Star Observers and the various Journals of the British Astronomical Association covering the entire period since its discovery.

The mean cycle of this star of 54.1 days is very like that of SS Cygni and although in this case too, there are deviations from the mean interval between outbursts, they are not quite as marked as in the case of the latter variable. The fluctuations at minimum also appear to be less pronounced although this may be due, in part, to the smaller number of observations which have been made visually during this phase.

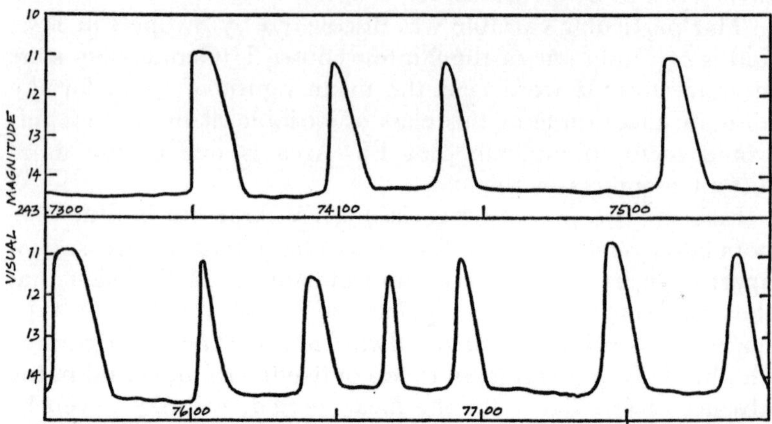

Fig. 28 – Light curve of SS Aurigae (*Courtesy of the British Astronomical Association*).

As the light curve shows (Fig. 28), the maxima are almost invariably characterised by an abrupt increase in brightness of about four magnitudes, the rise often taking place in a single day. So far, only two types of outbursts have been recorded – the classical long and short – both of which are sensibly smooth during the decline with the former being the brighter of the two.

From past records, there would seem to be a general tendency for the long and short maxima to alternate although, as with all of these variables, the maxima are quite unpredictable and there are occasional deviations from this rule. In northern latitudes, the star is circumpolar but during the summer months it lies very close to the horizon and is accessible only with difficulty. An 8-inch telescope, at least, is necessary to observe this variable when at minimum and then only in a clear sky with no moon.

Low-dispersion spectrograms of SS Aurigae were obtained in 1943 by Elvey and Babcock[8] using the 80-inch reflector at the McDonald Observatory, the dispersion being about

340 Å per millimetre. In common with most of the dwarf novae so far investigated spectroscopically, these show broad, bright emission lines of the Balmer series of hydrogen and the H and K lines of singly ionised calcium, together with the faint triplet of neutral helium, the latter also in emission. These bright lines are superimposed upon a continuum which has a distribution of energy similar to that of a G-type star. At maximum brightness, the underlying continuum has the characteristics of a type B star and is crossed by very faint absorption lines which are both broad and shallow.

A spectrogram taken by Herbig[16] with the Crossley reflector at Lick Observatory when SS Aurigae was at minimum shows rather wide emission lines of both hydrogen and calcium, with a very faint emission at λ4471 due to neutral helium. The underlying continuum shows very little extension into the ultra-violet.

A few high-dispersion spectrograms have been taken by Kraft[17] when the star was also at minimum, confirming the strong emission spectrum of hydrogen and also showing that the Balmer jump is in emission although long before the limit of the Balmer series is reached, the lines converge rapidly and blend into the continuum. From the radial velocities which have been measured, Kraft has also deduced an orbital period of about 3 hours and 30 minutes, a value which may require some revision once further observations are obtained although it is unlikely that any change in this value will prove to be very great.

The peculiar velocity of SS Aurigae has been found to be about +45 kilometres per second which is very similar to that of U Geminorum. The radial velocity of the blue star is about 85 kilometres per second; that of the cool component has not been measured owing to the difficulties encountered in observing the absorption lines of the late-type star in the photographic region of the spectrum. As we have already seen, those U Geminorum variables which have short orbital periods (less than about 6 hours) do not show the spectrum of the secondary since in these systems the cool companion is relatively faint.

At present, we cannot say whether SS Aurigae is an eclipsing system or not and no figures are available for its orbital inclination.

UZ BOÖTIS

$14^h\ 41^m\ 42^s\ +22°\ 23'.3$ (1950.0) $13^m.2–16^m.7$ (photographic)

This U Geminorum variable has been described by Hanley and Shapley[18] but it has not been systematically observed and even such a basic feature as its mean period is not known with any accuracy. Fortunately, as far as visual observation is concerned, it lies in a sparsely populated field and, apart from its faintness, presents few problems in the way of identification, especially when at, or near, maximum brightness.

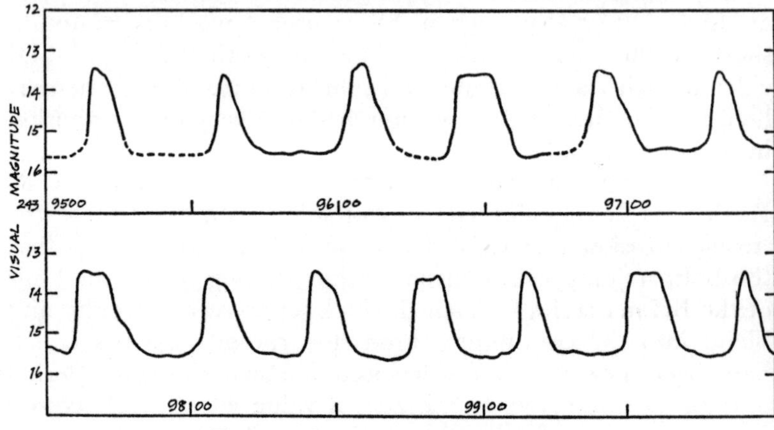

Fig. 29 – Light curve of UZ Boötis.

The light curve (Fig. 29), represents a total of 194 observations made by the author during 1967 and would suggest a mean period of about 40 days. However, since the variable was at the very limit of detection in the 13-inch reflector used throughout most of the minima, and the period covered by the investigation relatively short, this figure must necessarily be regarded as only tentative. Doubtless a more accurate mean cycle will be derived as more extended series of observations become available.

Although both long and short maxima were recorded during the period 1967–68, very few of the latter were observed. The rise to maximum is generally steep, the decline often showing pronounced humps. Any deviations from the mean period appear to be very small.

AF CAMELOPARDALIS

$03^h\ 28^m\ 18^s\ +58°\ 31'.5$ (1950.0) $13^m.4$–($17^m.0$ (photographic)

This faint member of the U Geminorum class lies just within the border of the Milky Way, the field being readily found from its proximity to the optical double Σ396 Camelopardalis. There are, unfortunately, a large number of faint stars in the same field and being so faint at minimum we know nothing at all of the behaviour during this phase. Comparatively few maxima have been recorded making it difficult to assign a mean period to this star. These outbursts which have been observed[7] all appear to have been typical of the dwarf novae.

The variable is circumpolar in northern latitudes but large instruments are required to observe this star satisfactorily. At present, no details of its spectroscopic characteristics are known.

EP CARINAE

$10^h\ 26^m\ 34^s\ -58°\ 42'.5$ (1950.0) 13^m–($16^m.2$ (photographic)

This peculiar variable was discovered by Hertzsprung[19] as long ago as 1928. It is visible on a small number of photographic plates and from the abruptness of its appearance and the apparently fairly large amplitude, it has been tentatively assigned to the U Geminorum class. Very little is known either of its light variations during a typical maximum, nor of its mean period. It is therefore quite possible that it is not a dwarf nova at all. Lying as it does in a very crowded field of fairly bright stars, we are again faced with difficulties of identification, even when at maximum. A long and detailed series of observations are required before EP Carinae can be unambiguously placed in the U Geminorum group.

The magnitude at minimum is not yet known although it probably lies beyond the capabilities of present spectroscopic techniques which might otherwise be used to determine its true nature.

DK CASSIOPEIAE

$00^h\ 15^m\ 29^s\ +57°\ 09'.2$ (1950.0) $15^m.4$–($17^m.4$ (photographic)

DK Cassiopeiae is one of four extremely faint U Geminorum

variables discovered in this particular constellation by that great variable star observer, Hoffmeister[20] in 1942–43. Because of its general faintness, the number of observations available for discussion are very few but there would appear to be little doubt that it is a member of the U Geminorum class. As might be expected, neither its mean period nor any detailed light curve have, as yet, been obtained. The field may be readily found from β Cassiopeiae, but as with all of the dwarf novae which lie within the boundaries of the Milky Way, there are several faint companions and very large apertures are needed to follow this star, even through one of its outbursts. It seems unlikely that any spectrograms will be obtained without further refinements in spectroscopic techniques.

FI CASSIOPEIAE

$00^h\ 03^m\ 22^s$ $+55°\ 38'.7$ (1950.0) $15^m.0$–($17^m.2$ (photographic)

Discovered on the same series of plates as the previous variable, this star is equally faint and so far has been observed only when at maximum. It is not surprising, therefore, that our knowledge of its light variations is extremely meagre and no spectrograms have been obtained at any phase of the light curve. As with the other very faint U Geminorum variables, this star would seem to be an extremely remote object even when allowance is made for the range of absolute magnitudes of these stars at minimum. Like DK Cassiopeiae, it is visible only in very large instruments.

GX CASSIOPEIAE

$00^h\ 43^m\ 59^s$ $+56°\ 36'.4$ (1950.0) $13^m.9$–($17^m.5$ (photographic)

If we assume an average amplitude of four magnitudes for the U Geminorum stars as a whole, the minimum for this star is probably not far below the limiting photographic magnitude of the plates taken by Hoffmeister[20] during the survey which resulted in its discovery. At maximum it is more than a magnitude brighter than either of the two preceding variables but in spite of this it has not been comprehensively observed, either visually or photographically, and consequently its mean period is unknown. Its spectrum, too, has not been recorded. From the observations made by Hoffmeister[20], however, there would seem little doubt that GX Cassiopeiae is a true member of the U Geminorum class.

HT CASSIOPEIAE

$01^h\ 06^m\ 59^s$ $+59°\ 48'.1$ (1950.0) $13^m.5$–$16^m.4$ (photographic)

This star is the brightest of the four U Geminorum variables found by Hoffmeister[20] in Cassiopeia at all phases of its light variation and is just within reach of a 13-inch telescope when seeing conditions are good and there is no moon. The photographic amplitude of only 2.9 magnitudes is quite small for this type of variable and the visual range is somewhat smaller. If we use the cycle-amplitude relationship to make an estimate of its mean period, we obtain a figure of approximately 30 days for this star.

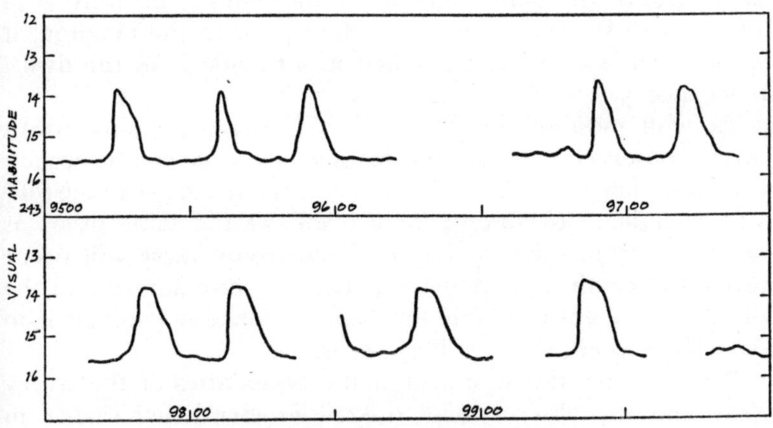

Fig. 30 – Light curve of HT Cassiopeiae.

How closely does this agree with observation? The light curve (Fig. 30), was obtained by the author during 1967–68 and is compiled from a total of 227 observations. Inevitably there are gaps during which an outburst may have occurred but it will be seen that on three occasions, two maxima were observed within a period of 30 to 35 days. This suggests (although it certainly does not prove) that the mean cycle length is of this order. Owing to the unpredictable nature of the outbursts of these variables, this figure must be regarded with caution until further observational data, covering a much longer period, become available.

The spectrum of this variable has not been recorded. Quite clearly, at minimum, it is doubtful if high-dispersion spectrograms can yet be obtained.

KP CASSIOPEIAE

$00^h\ 35^m\ 41^s$ $+60°\ 57'.5$ (1950.0) $14^m.0$–($16^m.5$ (photographic)

In 1948–49, Hoffmeister[21] carried out a further photographic survey of this particular constellation, a survey which resulted in the discovery of many more variable stars of all types. Among these were four faint U Geminorum stars; KP Cassiopeiae and the three which follow. So far, we have no light curve for this variable but since the star is clearly visible on several of the plates and below the limiting magnitude of 16.5 on others taken only a few days prior to the maxima, it would seem that we are justified in assigning it to the dwarf nova class.

As with most of the fainter dwarf novae, we must await further observational data on this star before we can be certain that it is not a Z Camelopardalis variable. Any 'standstills' will be difficult to observe for if it follows the same trend as the majority of stars in this small subgroup, these will occur around fifteenth magnitude or perhaps a little fainter and the number of plates available for examination is not sufficient to determine whether or not they occur.

The variable lies well within the boundaries of the Milky Way and the observational difficulties mentioned earlier in connection with such stars also hold in this case.

KU CASSIOPEIAE

$01^h\ 27^m\ 50^s$ $+57°\ 38'.6$ (1950.0) $13^m.0$–($17^m.5$ (photographic)

This U Geminorum variable has been a little more completely observed than the preceding star. Like it, KU Cassiopeiae was discovered photographically by Hoffmeister[21] and being a magnitude brighter when at maximum, it is more easily identified.

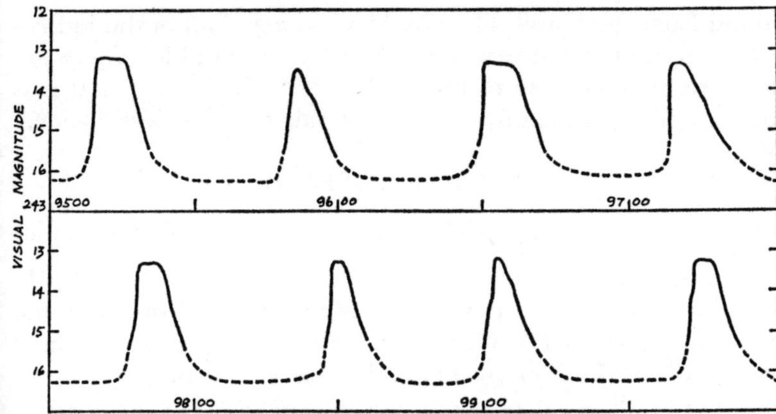

Fig. 31 – Light curve of KU Cassiopeiae.

A mean period of 60 days was assigned to KU Cassiopeiae by its discoverer and from observations made by the author during 1967–68 (Fig. 31), the average cycle throughout this period would seem to be one of about 63 days. During the two years when the variable was under observation, the outbursts were found to occur very close to this mean period and were of the usual two types, long and short, the former being flat at maximum. Marked irregularities are apparent on both ascending and descending branches of the light curve but nothing is known of the minima which were beyond the reach of the instrument used (the dotted portions of the light curve indicate that the variable was invisible at these times).

The spectrum of this star, like those of most of the fainter dwarf novae, has not yet been recorded and when at minimum it is certainly beyond the detection of present spectroscopic instruments. The star lies in the same field as the 4.88 magnitude star χ Cassiopeiae.

KZ CASSIOPEIAE

$23^h\ 05^m\ 51^s\ +56°\ 11'.2$ (1950.0) $14^m.5$–($17^m.0$ (photographic)

This faint dwarf nova lies in a densely populated field and no details have been published of either its light curve or spectrum. From the rapidity of its rise and the amplitude, as found from the plates taken by Hoffmeister[21], all of the indications are that this star is a U Geminorum variable. There are few bright stars close to the field of KZ Cassiopeiae and it is best found from the fifth magnitude pair 1 and 2 Cassiopeiae.

LM CASSIOPEIAE

$23^h\ 10^m\ 29^s\ +56°\ 35'.3$ (1950.0) $15^m.0$–($17^m.0$ (photographic)

LM Cassiopeiae is one of the faintest members of the U Geminorum class yet discovered and was found on the same series of plates as KZ Cassiopeiae. Not surprisingly, we know very little indeed of its light changes or mean period and nothing at all of its spectrum.

Clearly it is not a star which lends itself readily to visual observation and a long series of photographic estimates will be necessary for the determination of its light cycle and particularly of the brightness at minimum. Since most of the larger instruments in the major observatories of the world are committed to long programmes of investigation for which they alone are suited, it seems unlikely that further information will be forthcoming for many years.

BV CENTAURI

$13^h\ 28^m\ 10^s\ -54°\ 43'.0$ (1950.0) $10^m.4$–$14^m.0$ (photographic)

This variable is among the brightest of the southern U Geminorum stars and was discovered by Hoffleit[22] in 1930. Although the star has not been as fully observed as VW Hydri, several maxima have been well followed, particularly by the Variable Star Section of the Royal Astronomical Society of New Zealand[23]. The mean period would appear to be rather long, possibly about 80 days, although no definite figure has, as yet, been given.

From an examination of the light curve (Fig. 32), it will readily be seen that the maxima of this variable are very peculiar for a U Geminorum star. Unlike the majority of the dwarf novae, the rise is seldom steep. Indeed, it is very often more protracted than the eventual decline to minimum and a common feature is a pronounced pause at about magnitude 12.0 before the final rise ensues. In certain respects, therefore,

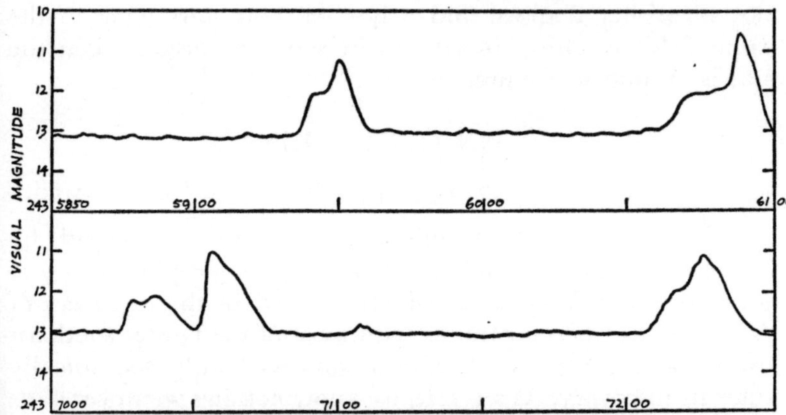

Fig. 32 – Light curve of BV Centauri (*Courtesy of the Variable Star Section of the Royal Astronomical Society of New Zealand*).

this variable resembles RU Pegasi which, as we shall see, also exhibits a fairly slow rise to maximum.

Few details of the spectrum of this star have been reported. There are, possibly, certain difficulties associated with any spectroscopic observations, especially when the star is at minimum. There are several stars of approximately the same magnitude quite close to the star which, although not too troublesome when it comes to making visual estimates of brightness, introduce difficulties for spectroscopic work.

MU CENTAURI

$12^h\ 10^m\ 18^s$ $-44°\ 10'.7$ (1950.0) $12^m.4$–$15^m.0$ (photographic)

Discovered when at maximum on May 11, 1931, MU Centauri has been described by Rybka[24] using observations made at Lwow Observatory. The observational data available, however, is not sufficiently complete for a mean cycle length to be established and although the star is well within the reach of present-day spectroscopic facilities, even when at minimum, few such investigations have been carried out. We must not lose sight of the fact, however, that this variable lies too far south to be satisfactorily observed by the great reflectors at Mount Wilson and Mount Palomar. As with most of the dwarf novae within this sprawling southern constellation, it is not surprising that we find a host of faint stars lying in the same field, resulting inevitably in some confusion when the star is around minimum.

NN CENTAURI

$13^h\ 11^m\ 04^s$ $-60°\ 36'.6$ (1950.0) $13^m.2$–$16^m.9$ (photographic)

This particular U Geminorum variable was found by Shapley and Swope[33] during the same photographic survey of selected Milky Way fields which resulted in the discovery of AT Arae. But here again, as with most of the fainter southern dwarf novae, the star has been observed only sporadically since its discovery. As a result, we know nothing whatever of its mean period or spectrum.

V373 CENTAURI

$12^h\ 23^m\ 21^s$ $-45°\ 32'.6$ (1950.0) $13^m.3$–$15^m.8$ (photographic)

V373 Centauri is a further faint member of the U Geminorum class which has not been seriously studied since it was found by Opalski[25] at the Lwow Observatory in 1935. For this reason, its mean cycle length is still uncertain. Observation of this star is made more difficult by the presence of at least three faint companions lying very close to the position given for the dwarf nova. The few maxima which have been satisfactorily observed all appear to have been of the characteristic abrupt type with a rapid rise to maximum and from this fact alone there would seem to be some justification for including it in the U Geminorum group although clearly much more observational information is desirable, particularly as the star belongs to that group of the dwarf novae which have quite small amplitudes.

There are other possibilities which must also be considered, of course. A light variation such as that which has been found for V373 Centauri could equally fit a variable of the RW Aurigae type but this does not seem likely. The maximum brightness attained at each outburst is fairly constant whereas the RW Aurigae variables are completely irregular, both in period and amplitude. The variable lies well away from the Milky Way and there are few bright stars in its vicinity to aid in locating the field. It is perhaps best found from γ and τ Centauri.

V436 CENTAURI

$11^h\ 12^m\ 13^s\ -37°\ 25'.9$ (1950.0) $11^m.9$–$15^m.4$ (photographic)

Discovered by Erro[26] in 1940, this variable is well removed from the Milky Way, lying on the borders with Hydra and Antlia. It is quite bright at maximum but again, very little is known of its light changes due to a paucity of observations. If we fall back upon the cycle-amplitude relationship, this would suggest a mean period of about 35 days for this star but insufficient data have been secured at present to confirm this. The maxima, at least, are well within the range of moderate-sized instruments and being so far from the Milky Way, the field is sufficiently open for identification at maximum to be quite simple.

V442 CENTAURI

$11^h\ 22^m\ 24^s\ -35°\ 37'.2$ (1950.0) $12^m.1$–($16^m.5$ (photographic)

V442 Centauri is a second U Geminorum variable found by Erro[26] during a photographic survey of the variable stars in this particular field. A mean period of 50 days has been given for this star, which taking into account the uncertainty about its minimum magnitude, would appear to be in fairly good agreement with that derived from the cycle-amplitude relationship for the dwarf novae. Accordingly, it is probable that any future modification of this figure, resulting from an extension of its light curve by additional observations, will be relatively minor. As with all of these variables, however, the light changes are quite unpredictable and it would be both unwise and imprudent to be dogmatic on this point.

Two nearby fifth magnitude stars make it comparatively simple to identify the field which lies close to the southern border with Hydra, some distance from the Milky Way in this region. There are also two eighth magnitude stars which aid identification still further. In spite of its relative brightness at maximum, the spectrum of this variable has not yet been described in detail.

V485 CENTAURI

$12^h\ 54^m\ 38^s$ $-32°\ 56'.3$ (1950.0) $13^m.0$–$(16^m.3$ (photographic)

An extensive survey of the northern boundary of Centaurus by Huruhata[27] in 1939–40 resulted in the discovery of a large number of variable stars, mostly RR Lyrae type variables. Among them, however, were two U Geminorum stars, V485 and V591 Centauri.

V485 Centauri, being the brighter of the two, has been the more closely studied and a very short mean period of about 10 days has been assigned to it. There is still some uncertainty about this period and further observational data are needed in order to confirm it or otherwise since, if found to be accurate, it is the shortest mean period of all the U Geminorum variables which have so far been investigated; indeed, it is more like that of a Z Camelopardalis star than one of this particular class although we do have a close counterpart in SU Ursae Majoris. In view of this, and bearing in mind that we still lack a light curve which is anything like complete even over a relatively short period, we should perhaps keep an open mind on this question.

The field contains several bright stars, including a close eighth magnitude companion. The variable itself forms the apex of a small triangle of stars and some difficulties are inevitable when visual estimates are made, particularly if averted vision has to be used when the variable is at the limit of vision.

V591 CENTAURI

$12^h\ 39^m\ 36^s$ $-33°\ 17'.3$ (1950.0) $14^m.0$–$16^m.0$ (photographic)

As mentioned above, this faint U Geminorum star was discovered on the same plates as V485 Centauri by Huruhata[27] and is a magnitude fainter at maximum. The magnitude at minimum is known, however, yielding the small amplitude of only two magnitudes. Very little has been published concerning the typical light variations and the spectrum has not been recorded.

Z CHAMAELEONTIS

$08^h\ 08^m\ 40^s$ $-76°\ 23'.5$ (1950.0) $11^m.4$–$15^m.3$ (visual)

The light variations of Z Chamaeleontis, as shown by a study of its light curve (Fig. 33), show that it is a fairly typical U Geminorum variable. Being one of the brighter southern dwarf novae, it has been well observed in recent years, particularly by the Variable Star Section of the Royal Astronomical Society of New Zealand[23].

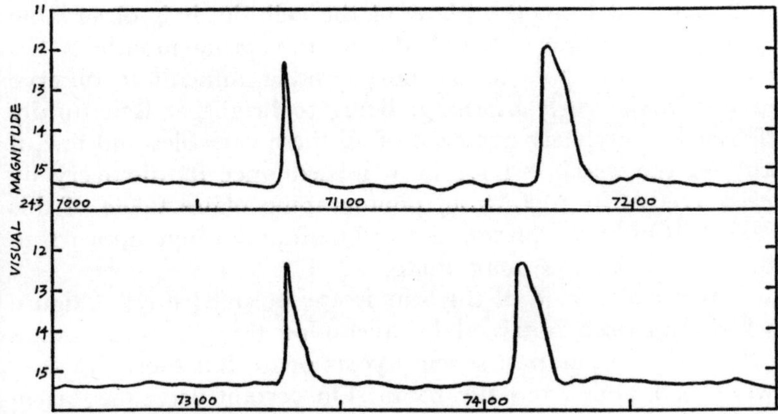

Fig. 33 – Light curve of Z Chamaeleontis (*Courtesy of the Variable Star Section of the Royal Astronomical Society of New Zealand*).

In common with most of the variables in this class, both long and short maxima occur, the former being usually about a magnitude brighter than the latter. The rise from minimum is quite steep, especially in the case of the short maxima, these outbursts being characterised by a stay at peak brightness of only about one day, the star then commencing to fade quite rapidly. During the longer maxima, the variable remains at its most brilliant for between four and six days, the decline in this case, being somewhat slower. The minima are generally reasonably undisturbed (although naturally fewer observations are available of the star during this phase of the light variations) and any fluctuations are quite small, normally of the order of 0.3 magnitude. The light curve of this star shows some similarities to that of VW Hydri which will be discussed later in the chapter. The mean period of Z Chamaeleontis is about 70 days.

SS CYGNI

$21^{h}\ 40^{m}\ 44^{s}\ +43°\ 21'.3$ (1950.0) $8^{m}.1$–$12^{m}.1$ (visual)

As we have seen earlier, SS Cygni is the brightest of all the dwarf novae both at maximum and, if we discount the very peculiar and atypical AE Aquarii, at minimum also. Being a circumpolar star in northern latitudes above about +50° and well removed from the plane of the ecliptic, it is observable all the year round although during the spring months it lies well below the pole to the north and is difficult to observe except in the early morning. Being so bright, it is naturally the most completely observed of all these variables and few, if any, of the maxima have been missed since its discovery by Miss Wells[28] in 1896, from photographic plates taken of this region. On these plates, the photographic range appears to be from 7.6 to 11.8 magnitude.

The mean period of this star is one of 50.64 days, a figure which has been obtained by averaging the 500 or so cycles observed over the past seventy years or so, but there are very wide deviations from the mean. On certain occasions, as in 1960, twelve maxima have been recorded, the mean interval then being as short as 33 days, while at other times, the number of outbursts has been as low as five with a mean cycle length as long as 71 days. Apart from being the most easily observed dwarf nova, therefore, such behaviour also makes it one of the most interesting of these stars, since these wide variations from the mean are also reflected in the duration of the various minima. At times, SS Cygni has been known to remain at minimum for almost 100 days while in October 1959, shortly before its period of greatest activity, its stay at minimum lasted for only one day before a further slow rise began.

We have already seen in Chapter Two that the maxima of this star vary widely in shape, brightness and duration. During the early years of its investigation, these were divided into three general classes; long, short and anomalous, the last named being characterised by a relatively slow rise to maximum, a more rounded peak than either of the other two and, in general, a somewhat slower decline to minimum. Instances have been known, however, of anomalous maxima in which the decline is appreciably more rapid than the rise. In general, it may be said that where there are any marked irregularities during the maximum and in particular if the rise is protracted, the outburst is classed as anomalous. As a result, we find that this type of maxima form a very heterogenous class. There are often pronounced humps on both the ascending and descending branches of the light curve, particularly so in the case of the anomalous maxima and quite often, too, we find a slight dip in the minimum immediately preceding a rise. This simple classification is still in use by variable star observers, especially in this country.

A much more comprehensive system of classification is that due to Campbell (Fig. 34). Here, four general types are recognised, designated A, B, C and D based upon the steepness of the rise to maximum; Type A being the steepest and Type D the least rapid. Each type is further subdivided depending upon the number of days spent brighter than magnitude 10.0.

A type A9 maximum, for example, would indicate one showing a very rapid rise from minimum and remaining above tenth magnitude for 9 days. A D6 maximum, on the other hand, would remain brighter than magnitude 10.0 for only 6 days

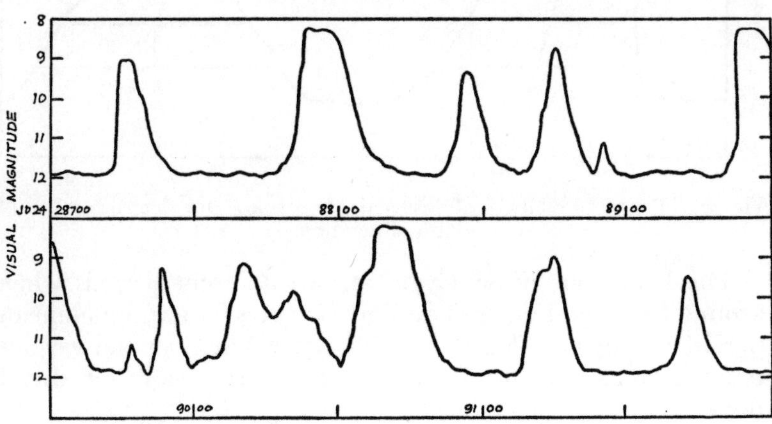

Fig. 34 – Typical light curve of SS Cygni (*Courtesy of the British Astronomical Association*).

and the rise would be comparatively slow. Here one might ask why magnitude 10.0 has been chosen as the reference point during the rise and decline and not the time actually spent at maximum brightness. The answer is really quite simple. In practice, it is found to be quite a simple matter to determine from the light curve the dates on which the star attains tenth magnitude on both the ascending and descending branches of the maximum, whereas when the star is around its peak brightness there are often minor fluctuations of a few tenths of a magnitude, making it far from easy to define an actual maximum brightness. The situation is very similar to that found when determining the dates of the maxima. Due to the irregularities which are often present in the light curve, it would be very difficult, if not impossible, to compute the actual date simply by inspection and the standard method used almost universally, is that of bisected chords, first put forward by Pogson (Fig. 35).

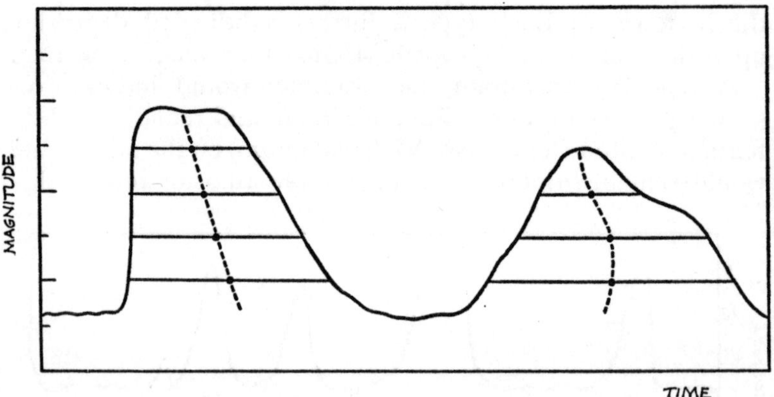

Fig. 35 – Pogson's method of bisected chords for the determination of dates of maxima.

The behaviour of SS Cygni at two different periods when around the maximum and minimum values for its annual mean period is shown in Fig. 36 which also exemplifies the various types of maxima which have been recorded for this dwarf nova.

Statistical analyses of the numbers of each types of maximum on Campbell's classification have been made by several astronomers with interesting results. Of the 504 maxima which have been observed up to the end of 1967, the author has assigned 62, 16, 12 and 10 per cent of types A, B, C and D respectively. Clearly, from this, the most common kind of maximum is that characterised by a steep, uninterrupted rise with the variable brightening by approximately four magnitudes in the space of a single day. An examination of Fig. 36 also shows us that even when the star is at minimum brightness, its behaviour is often far from regular. Even visual estimates which are, at best, only accurate to within 0.2 magnitude, are sufficient to indicate that quite marked fluctuations in brightness are common during this phase and these will be discussed later. The magnitude at minimum, too, is far from constant. Some are known in which the variable has been as

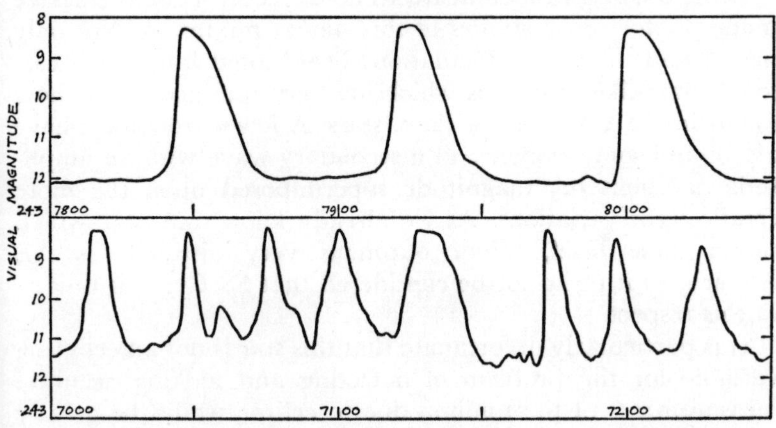

Fig. 36 – The light variations of SS Cygni at different values of its mean annual period (*Courtesy of the British Astronomical Association*).

faint as 12.1 magnitude while at the other end of the scale there have been some years when the star has not fallen below magnitude 11.5. In addition, a small number of well-authenticated subsidiary rises have been recorded during which the variable has become no brighter than tenth magnitude. Such subsidiary maxima are not normally included in the numbered sequence for this dwarf nova.

In Chapter Two, we also discussed the statistical analysis which was made by Lortet-Zuckermann[29] which would indicate that the outbursts of SS Cygni can, in part, be explained on the basis of a simple Markov chain. There are, unfortunately, several difficulties in applying this theory even to this star which is the most thoroughly observed of all the dwarf nova. When Markov chains of higher orders are considered, it is found that these do not satisfactorily explain the existence of those periods in the star's light variation when many minor outbursts are recorded, nor when there are large numbers of normal maxima such as occurred in 1960. At present, it does not seem as though this problem can be readily solved by this means. Certainly it would appear to be of a far more complex nature than was originally thought. So far, there is little real observational evidence to suggest that any particular outburst is directly influenced by those immediately preceding it.

Being the brightest of the dwarf novae, several observers have made photoelectric studies of this star at minimum. Not only have small, irregular fluctuations been noted but also occasional, flare-like eruptions which are very reminiscent of those found in the UV Ceti, or flare, stars. A few astronomers have also found some evidence of a secondary wave with an amplitude of about 0.5 magnitude superimposed upon the more erratic light variations. As we already know, one star which shows these brief, minor outbursts very distinctly is AE Aquarii, so it must not be considered that SS Cygni is unique in this respect.

It is particularly unfortunate that this star is not an eclipsing variable for the problem of detecting and making accurate measurements of the minima due to eclipse would be greatly simplified in a system which is as bright as SS Cygni at minimum. Instead, as we shall see later, the two systems which do show evidence for eclipse, U Geminorum and Z Camelopardalis, have minima as faint as 14.5 magnitude and 13.1 magnitude respectively.

Let us now consider some of the spectroscopic changes of SS Cygni (these have been discussed in detail earlier in the book). Because of its brightness at all phases of its light variation, this variable is the easiest of all the dwarf novae to observe spectroscopically throughout the whole of its light cycle. In addition, it is one of only three of the dwarf novae (the other two being EY Cygni and RU Pegasi) which show the absorption spectrum of the late-type companion in the photographic region. As the variable begins a rise to maximum, bright emission lines of hydrogen and singly ionised calcium appear which, however, become progressively weaker as the star brightens still further, there being a corresponding increase in the intensity of the absorption lines of the Balmer series of hydrogen. For a relatively short period during the rise, when the star is at about tenth magnitude, double absorption lines appear which gradually coalesce and widen until at maximum the spectrum is dominated by broad, shallow absorption features. As the variable fades following a maximum, the sequence of events is, of course, reversed.

We find that the colour of SS Cygni also varies throughout a typical maximum. At minimum, the star possesses a yellow colour. As it brightens, however, the colour becomes appreciably bluer, the cause being a shift of the peak intensity in the underlying continuum towards the violet end of the spectrum.

As with all of the dwarf novae which are within reach of spectroscopic examination, the blue component of SS Cygni is an sdBe-type dwarf, the secondary being a red dwarf of spectral class dG5 with an estimated mass somewhat less than that of the sun. Since the variable does not eclipse, we must resort to indirect measures for the determination of the masses of the individual components, hence the various figures which have been published for the masses. There is one curious anomaly associated with the red star which deserves special mention here, namely that it is about three magnitudes underluminous for its estimated mass and spectral type. One reason for this peculiarity is, as we have seen, that for various reasons there is a tendency to over-estimate the spectral class of these secondaries, another being that mass is lost from this star through the inner Lagrangian point.

The parallax of SS Cygni has been fairly well established by Strand[30], yielding an absolute magnitude at minimum of +9.5 and a distance of approximately 217 light years. The variable is therefore sufficiently close to us for there to be only a negligible amount of interstellar reddening and since, at maximum, the spectrum is almost continuous, we may assume that, to a first approximation, it radiates as a black body at this phase of its light cycle. This being so, the temperature of SS Cygni, at maximum, is of the order of 12,000°K, a figure which is in good agreement with an absolute magnitude of +9.5 at minimum and also with that derived for the surface temperatures of normal white dwarfs by independent method. However, such a result does not agree at all well with the presence of a star of spectral type dG5 and about 0.6 solar mass. Since the discovery by Krzeminski[31] that in the U Geminorum system, it is the red component which is responsible for the outbursts, this anomaly virtually disappears for we may assume (although in only this one particular case has it yet been proved) that the surface temperature of the red secondary of SS Cygni is also in the neighbourhood of 12,000°K during an outburst.

We must, however, not lose sight of the fact that this phenomenon of underluminosity is also found in the W Ursae Majoris variables, a feature which, together with their short orbital periods, has been advanced in support of the theory that the dwarf novae are generically related to these variables, possibly representing a later stage in the evolutionary history of the W Ursae Majoris variables.

EY CYGNI

$19^h\ 52^m\ 40^s$ $+32°\ 13'.7$ (1950.0) $11^m.4$–$15^m.7$ (photographic)

This dwarf nova has not been as extensively observed as the previous star for several reasons. Although it is, like the other, circumpolar in northern latitudes, EY Cygni lies on the very edge of the Milky Way in a much richer field of stars and is also about three magnitudes fainter than SS Cygni at all phases of its light cycle. The presence of several very faint companions makes identification troublesome around minimum. Unlike the previous star, the outbursts are much less frequent and the mean period would appear to be approximately 1,000 days which is extremely long for a star of the dwarf nova class.

The light curve (Fig. 37) illustrates a typical maximum exhibited by this variable which was quite fully observed by the American Association of Variable Star Observers[7] and fully confirms its assignment to the U Geminorum class.

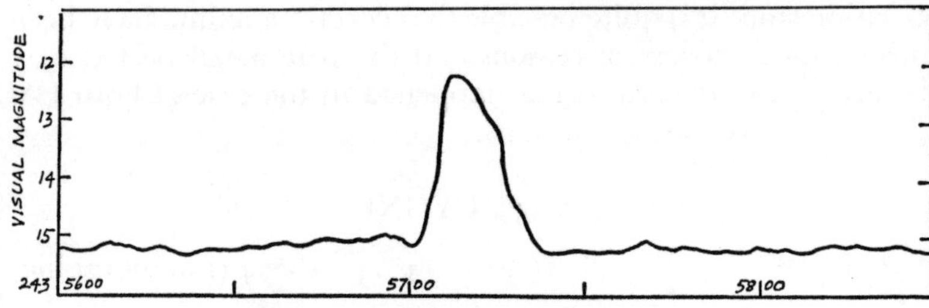

Fig. 37 – Light curve of EY Cygni.

Several high-dispersion spectrograms have been obtained around minimum, mostly by Kraft[17] during a spectroscopic study of these variables. This series of spectra, which were taken with the 200-inch reflector at Mount Palomar, shows that the absorption lines of a late-type dwarf are present at minimum but any velocity variation of these lines is extremely small, lying at the very limit of detection. This composite nature of the spectrum and the small velocity variation of the spectral lines provides us with evidence of a binary system for this star and would also suggest that we are viewing this system virtually pole-on. The late-type companion would seem to be of spectral type Ko and luminosity class V, somewhat later than the secondaries found in most of the dwarf novae. Although the orbital period is not yet known, the fact that the absorption spectrum of the secondary star is visible at minimum would indicate that it is no longer than 6 hours, for as we have seen, owing to the relationship between orbital period and absolute magnitude of the late-type companion, those with periods of less than 6 hours show no evidence of a composite spectrum. The emission lines in the spectrum of EY Cygni are extremely narrow, much more so than is normally found in these variables. This may well be a consequence of viewing this system pole-on since there will then be little axial rotating in the line of sight to result in a broadening of the lines. The peculiar radial velocity of this star is about +13 kilometres per second and the proper motion is 0.062 seconds of arc per year, the latter being the mean of the two values given by Mannino and Rosino[32] and Miczaika and Becker[33].

Owing to its apparently very long mean cycle, EY Cygni is a dwarf nova which should be observed on every possible occasion since it is quite possible that certain maxima have been missed for a variety of reasons and the true mean period may be eventually revised as has happened in the case of both UV Persei and SW Ursae Majoris.

V337 CYGNI

$19^h\ 58^m\ 06^s$ $+39°\ 05'.8$ (1950.0) $14^m.4$–($16^m.4$ (photographic)

We have in V337 Cygni one of the small group of variables which, although possibly members of the U Geminorum class, have been insufficiently observed for any firm conclusion to be reached concerning their real nature. The star was discovered in 1933 by Baade[34] and the only available observations of it have been made when at maximum and then only on a few isolated occasions. The star lies deep within the Milky Way and the field is best found from the fifth magnitude star 22 Cygni. Even at maximum, however, it is not an easy variable to observe visually. Large apertures are necessary and there are several close companions from which it must be clearly differentiated. When at its brightest, it is just within the reach of spectroscopic examination with a slit spectrograph using the largest instruments but so far its spectrum at this phase has not been reported. At minimum, of course, it is well beyond the capabilities of presently available equipment.

V503 CYGNI

$20^h\ 25^m\ 32^s$ $+43°\ 31'.7$ (1950.0) $13^m.4$–(16.7 (photographic)

V503 Cygni was discovered in 1949 by Hoffmeister[35] who deduced from the preliminary light curve a mean period of 30 days. This would suggest (although it certainly does not prove) that its minimum magnitude does not lie far below the photographic limit given above since, if it obeys the cycle-amplitude relationship, the expected amplitude would not be much more than three magnitudes.

The light curve given in Fig. 38 is derived from 379 observations made by the author during 1967–68, the dotted portions of the curve indicating that the variable was below the limiting magnitude of the 13-inch reflector used throughout this series of observations. The rise to maximum is generally

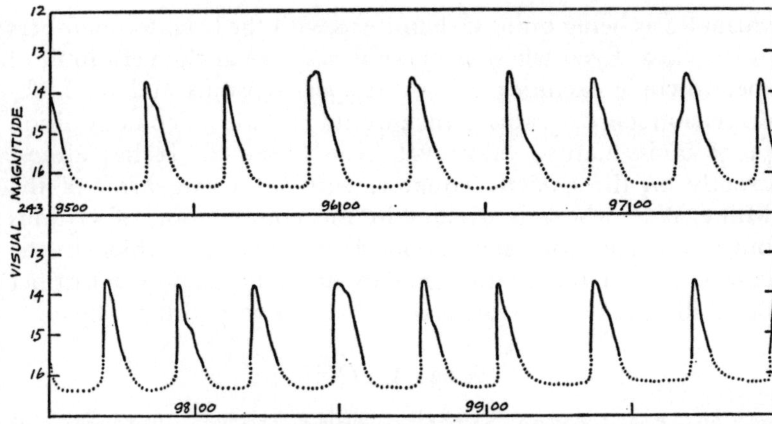

Fig. 38 – Light curve of V503 Cygni.

rapid and smooth although on occasion, there is a pronounced hump. Marked fluctuations are more common on the descending branch of the light curve and occur at almost any point during the decline. As will also be seen from Fig. 38, it is possible to divide the maxima into long and short types although the distinction is certainly not as pronounced as in the case of other dwarf novae and the long maxima do not appear to be as numerous as in the case of, say, SS Cygni or U Geminorum.

Very little is known of the behaviour of V503 Cygni when at minimum and the spectrum has not yet been described. The field, which is readily found from γ Cygni, lies in the Milky Way and is rich in moderately bright stars.

V550 CYGNI

$20^h\ 03^m\ 09^s\ +32°\ 13'.0$ (1950.0) $15^m.0$–($18^m.0$ (photographic)

V550 Cygni is one of the faintest of the U Geminorum variables yet discovered. The star was discovered on plates taken by Hoffmeister[21] and although visible on only a small number of these, its abrupt rise and the general characteristic of the light curve throughout a maximum are sufficient to assign it to the dwarf nova class. Naturally, since the amount of information we have concerning its light variations is still very small, we cannot regard its inclusion among the U Geminorum variables as being quite as definite as with the brighter members of the class. Even when at maximum it lies at the very limit of spectroscopic facilities using slit spectrographs and no high-dispersion spectrograms (which would provide welcome addition proof of its nature) have yet been obtained. It lies almost exactly on the galactic equator, within the two arms of the Milky Way where it divides in the constellation of Cygnus but nevertheless, the field abounds in faint stars which in the case of visual observation makes misidentification a distinct possibility. Like FV Arae, it is probably a very distant object.

V630 CYGNI

$21^h\ 32^m\ 59^s$ $+40°\ 26'.3$ (1950.0) $13^m.6$–($16^m.3$ (photographic)

This variable is one of two U Geminorum variables discovered in 1950 by Rohlfs[36]. It lies in the same field as V632 Cygni described below, very close to the fifth magnitude star 74 Cygni, not far removed from the borders of the Milky Way. V630 Cygni is not an easy star to observe even when at maximum since there are several faint stars of approximately the same magnitude very close to its position and high magnifications are necessary to separate these stars sufficiently for accurate visual estimates to be made. Neither the mean periods nor any spectroscopic details are known for this variable.

V632 CYGNI

$21^h\ 33^m\ 54^s$ $+40°\ 12'.3$ (1950.0) $12^m.8$–$16^m.2$ (photographic)

The second dwarf nova discovered in this field by Rohlfs[36], V632 Cygni, is reasonably bright at maximum; from the light curve (Fig. 39), the visual range as determined by the author would appear to be 13.1 to 15.6 magnitude although it must be borne in mind that these figures are based upon a relatively small number of observations covering only a short period and further data are desirable before these can be firmly established.

The mean cycle length over the period during which it has been observed is about 55 days. In general, the long and short maxima tend to alternate but on two occasions, an anomalous maximum was observed, characterised by a slow

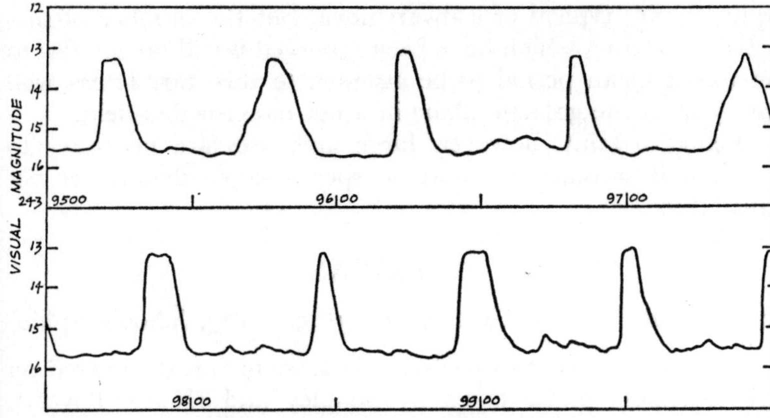

Fig. 39 – Light curve of V632 Cygni.

rise and a more rapid decline with very marked fluctuations on both ascending and descending branches of the light curve, a behaviour very reminiscent of SS Cygni. The minima, too, are marked by frequent and relatively large variations in brightness at irregular intervals.

V750 CYGNI

$20^h\ 47^m\ 45^s\ +50°\ 21'.1$ (1950.0) $13^m.4$–($16^m.5$ (photographic)

Although this U Geminorum variable is again one situated within the Milky Way, the field is fairly open in contrast to most other dwarf novae lying close to the galactic plane. There are, however, three faint companions nearby, one of which is sufficiently close to the variable for some confusion to be inevitable whenever the latter is around minimum. Very little is known, as yet, of its mean period and no spectrograms have been obtained.

XZ ERIDANI

$04^h\ 09^m\ 10^s\ -15°\ 31'.5$ (1950.0) $14^m.6$–($16^m.5$ (photographic)

This particular U Geminorum star was found by Shapley and Hughes-Boyce[37] in 1933 during a photographic search of this constellation for variable stars. Judging from the photographic evidence, the rise to maximum would appear to be quite rapid, typical of a dwarf nova, but the number of successive maxima which have been observed is still far too few to enable a mean period to be assigned to this star. It lies well away from the galactic plane in a not too crowded field.

Being so faint, however, large apertures are necessary to observe it satisfactorily and no spectroscopic details are yet available.

AH ERIDANI

$04^h\ 20^m\ 23^s\ -13°\ 28'.6$ (1950.0) $13^m.5$–$(16^m.5$ (photographic)

Like the preceding variable, AH Eridani was discovered on photographic plates taken by Shapley and Hughes-Boyce[37]. It, too, lies in a fairly sparsely populated field but like so many of these fainter U Geminorum variables it has not been closely observed since its discovery and few details of its light and spectroscopic characteristics are available for discussion. Since its magnitude at minimum has not yet been accurately defined, we cannot use the cycle-amplitude relationship even to make a reasoned guess at the mean cycle length.

AQ ERIDANI

$05^h\ 03^m\ 44^s\ -04°\ 11'.9$ (1950.0) $12^m.5$–$16^m.5$ (photographic)

AQ Eridani, discovered in 1935 by Hoppe[38], is readily located from its proximity to the third magnitude star β Eridani. There are three fairly bright stars in the same high-power field as the variable and a close, faint companion from which it has to be clearly separated when the variable is around minimum. In this respect, it closely resembles RU Pegasi.

Since the variable lies reasonably close to the plane of the ecliptic, it is unobservable for about three months of the year during the spring and early summer and several maxima are unavoidably missed when the sun is passing through the southern region of Taurus. Although it lies just south of the celestial equator, it is reasonably well situated for observers in the British Isles, particularly when near the meridian.

The light curve, as determined by the author (Fig. 40), suggests a mean period of approximately 82 days which, taking into consideration the relatively short period during which the

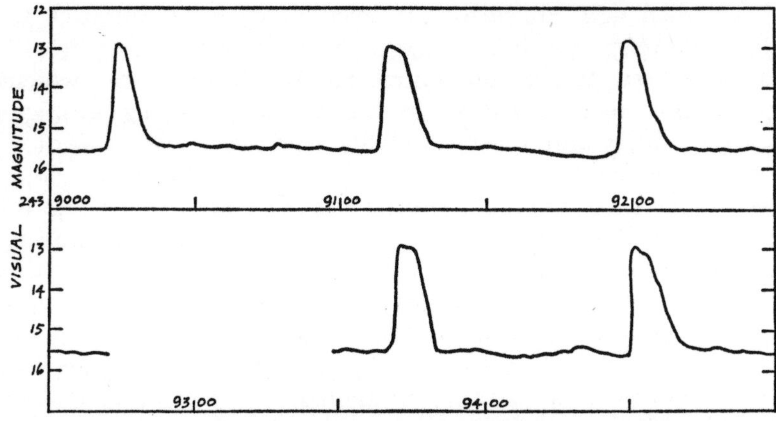

Fig. 40 – Light curve of AQ Eridani.

star was observed, is in quite good agreement with that of about 78 days given by Hoppe. From the above light curve, the visual range would appear to lie between 12.3 and 15.7 magnitude. All of the maxima so far observed have been of the abrupt type, one or two having irregularities on the descending branch. The minima are fairly flat with only minor fluctuations. Any deviations from the mean period seem to be quite small but further observational data are necessary before a true picture can be obtained. No high-dispersion spectrograms of this variable have been described.

U GEMINORUM

$07^h\ 52^m\ 08^s\ +22°\ 08'.3$ (1950.0) $8^m.8–14^m.4$ (visual)

As we have seen earlier, U Geminorum was the first of the dwarf novae to be discovered, by Hind[39,40] in 1855 when it was initially thought to be a nova. This view was shown to be erroneous and its true nature as a new type of variable star, recognised only when it was reobserved at maximum some three months later by Pogson[41]. It is unfortunate that the star lies almost on the ecliptic, since this results in several maxima being lost completely owing to its proximity to the sun for about three months of the year from the beginning of May to the middle of August. In spite of this, the star has been assiduously observed every year since its discovery.

Detailed light curves of the maxima from March 1856 to January 1907 have been given by Van der Bilt[42] where it will be seen that, making due allowance for those maxima which could not be observed when the sun was passing through Gemini, long and short outbursts tend to alternate with a remarkable constancy. From the vast number of observations which have accumulated during the past century or so, we now know that this alternation of long and short maxima is not an invariable rule, instances being recorded where two longs or two shorts have followed in succession. The long type of maximum is usually about half a magnitude or more brighter than the short and in every outburst so far observed, the rise has been of the rapid type; no anomalous maxima have ever been positively recorded.

The observations made of U Geminorum during the latter half of the nineteenth century suffer from the disadvantage that although the observers themselves were competent in the methods of variable star observation, they used different comparison stars in estimating the brightness of the variable. Even when the same stars were used, the magnitudes assigned to them by different investigators were often at variance. Some observers even changed their adopted magnitudes at various times without any consistency behind such changes, thereby adding to the difficulties of drawing accurate light curves. As a result, both the shape of the maxima and the epochs lack uniformity. Nevertheless, in spite of this somewhat unsatisfactory situation, the record they provide must not be despised since it extends our knowledge of the peculiar light variations of this star by almost half a century. A further difficulty, of course, is that few of the early investigators possessed sufficiently large instruments to see the variable when at minimum and this phase was almost totally unobserved.

It is worth mentioning at this point that some of the early observers reported peculiar light fluctuations of U Geminorum when around maximum brightness. Hind described the star when first discovered as having a curiously planetary appearance and according to Pogson[41], variations amounting to as much as four magnitudes in a period of between six and fifteen seconds were observed. Other equally experienced observers saw no such light fluctuations of this magnitude and none have been recorded in later years. It is now generally accepted that these apparently violent fluctuations were spurious, possibly being an optical effect.

The mean period of this star, averaged over many decades, is one of 102.96 days. It is therefore one of the longer period U Geminorum variables but as in the case of SS Cygni, there are very wide annual variations from the mean, the range being from 62 to 137 days. Owing to the difficulty of observing this star at certain times of the year, there is naturally an intrinsic uncertainty associated with these values, particularly the latter figure which may be possibly even higher than that given.

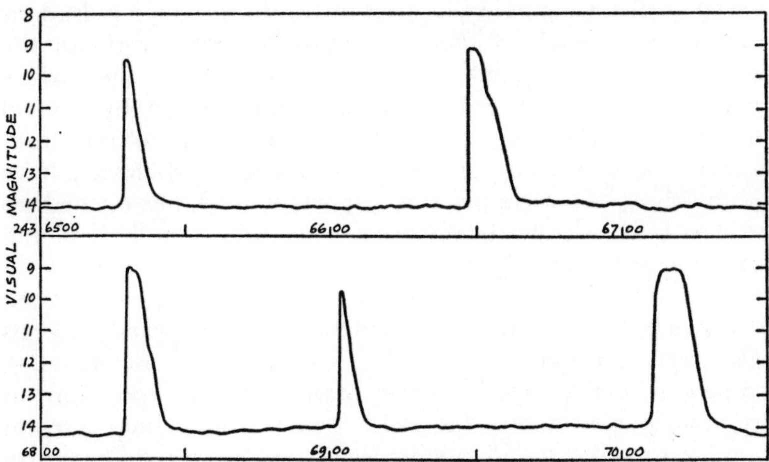

Fig. 41 – Light curve of U Geminorum (*Courtesy of the British Astronomical Association*).

The light curve of U Geminorum, illustrating the two types of maxima, is given in Fig. 41. As has already been mentioned, the long maxima are usually about half a magnitude the brighter of the two.

In Chapter Two, we saw how the light variations of the dwarf novae are of great importance when the variables are at minimum and this is especially so in the case of U Geminorum for, since the inclination of the orbit is very close to 90°, there is now definite evidence of an eclipse in this system. Extremely sensitive photoelectric equipment is, of course, necessary for the light fluctuations during the eclipse to be accurately measured. There are, as we have seen, many problems associated with the complex light changes during eclipse, particularly when we try to analyse the results obtained by three-colour photometry. Up to a few years ago, it seemed virtually impossible even to analyse the data which had been obtained.

The big breakthrough in our knowledge of the physical nature of this star, and in particular of the outbursts, has come with the discovery by Krzeminski[31] that the cool, red dwarf is responsible for the sudden brightenings. The physical mechanism by which this occurs has already been discussed in detail earlier where we saw that owing to irregular pulsations in the outer levels of this star, mass is ejected through the inner Lagrangian point while at the same time, the surface temperature of the secondary increases sharply by several thousand degrees. We must not imagine that all of the problems associated with this, and the other dwarf novae, have been solved but certainly this important work has opened up several fresh avenues of research which will be fully explored in the near future.

The spectroscopic features of U Geminorum have been investigated by many astronomers over the years since its discovery. As long ago as 1882, Copeland[43] and Knott[44], using an ocular spectroscope, described the spectrum as essentially continuous but with some bright lines present similar to those found previously in the novae. Some thirty years later, Miss Fleming[45] obtained objective prism spectra of both U Geminorum and SS Cygni and pointed out the highly peculiar nature of the spectral features, many of which could not be resolved owing to the low dispersion used.

When at minimum, the spectroscopic features, particularly the Balmer emission lines are as strong as those displayed by SS Cygni and Elvey and Babcock[8] also noted a bright Balmer continuum together with the presence of the stronger lines of neutral helium and the K lines of singly ionised calcium. The intensity of the latter is much weaker than in the minimum spectrum of SS Cygni and from this it may be inferred that the G-type companion (whose absorption spectrum has not yet been recorded in the photographic region) is as bright as, or perhaps a little brighter than, that of SS Cygni relative to the blue star. The true visual range of this variable may therefore be as great as six magnitudes.

Kraft[17] has obtained high-dispersion spectrograms of this star at minimum brightness and shown that the emission lines are conspicuously doubled. From the radial velocity curve, an orbital period of 4 hours and 10.5 minutes has been derived. The eccentricity of the orbit is 0.05, implying that it is essentially circular.

These high-dispersion spectrograms have also shown that the blue component, together with the brighter part of the rotating ring of gas surrounding it, is the object which is eclipsed at the time of primary minimum. No secondary minimum has been detected even when the highest gains compatible with the optimum signal-to-noise ratio were employed.

The field of U Geminorum is readily found from κ Geminorum and when at maximum can scarcely be missed owing to the paucity of bright stars in the vicinity.

UV GEMINORUM

$06^h\ 35^m\ 48^s$ $+18°\ 18'.9$ (1950.0) $14^m.7$–($17^m.2$ (photographic)

In spite of its extreme faintness, a mean period of 58 days has been assigned to this U Geminorum variable by Hoffmeister[46] from estimates made from a series of photographic plates taken of this region. However, the interval over which it has been anything like continuously observed is so small, and the difficulties associated with its observation such, that this mean period will, in all probability, have to be revised somewhat as more observations are accumulated.

AW GEMINORUM

$07^h\ 19^m\ 33^s$ $+28°\ 36'.2$ (1950.0) $13^m.0$–($17^m.4$ (photographic)

This U Geminorum star lies in a fairly densely populated field of faint stars close to the bright pair, 59 and 60 Geminorum. The number of available observations is still insufficient for a complete light curve to be drawn and we have very little reliable information about its mean cycle length.

The maxima are well within the reach of a 10-inch reflector although the minima are, of course, beyond the capabilities of anything but the largest instruments.

CH HERCULIS

$18^h 32^m 43^s$ $+24° 45'.5$ (1950.0) $13^m.5$–$17^m.2$ (photographic)

CH Herculis was found by Ahnert[47] on plates taken in 1940–41 and lies in a field containing several faint stars which are of a similar brightness to the variable when at maximum making positive identification extremely difficult even at this phase. The mean period is not yet known although the observations made so far indicate that the star is a true U Geminorum variable. On the few occasions when it has been photographed at maximum, the rise would appear to have been of the steep, rapid variety characteristic of these stars.

PR HERCULIS

$18^h 06^m 28^s$ $+38° 47'.3$ (1950.0) $14^m.0$–($17^m.5$ (photographic)

This variable, and the following dwarf nova, were discovered by Hoffmeister[48] in 1950. From its general faintness, it is not surprising that it is not an easy star to observe visually, even at maximum, and no details are available concerning either its mean cycle length or its spectrum. Clearly, it is not one of the dwarf novae which is within the scope of most visual observers and is more suited to photographic techniques using long exposures; a method which, although of great value in the actual discovery of these stars, has its limitations when it comes to providing reasonably complete light curves.

PR Herculis is a circumpolar variable, lying on the border with Lyra in a field devoid of bright, naked-eye stars. During the summer months, it is very close to the northern horizon and is accessible only during the early morning. The field of the variable is situated almost midway between Vega and θ Herculis.

PU HERCULIS

$18^h\ 07^m\ 58^s\ +31^\circ\ 57'.4$ (1950.0) $15^m.8$–$(17^m.6$ (photographic)

A very faint, and presumably distant, U Geminorum variable, PU Herculis was also discovered by Hoffmeister[48]. The observational difficulties mentioned in connection with the last dwarf nova are even more acute in the case of this star owing to its faintness. Not surprisingly, our knowledge of its behaviour is extremely meagre. From its appearance on the plates taken in 1950, the rise to maximum is clearly very rapid and suggestive of a dwarf nova. The interval between successive observed maxima would seem to rule out the possibility of it being a short-period irregular variable of the RW Aurigae class.

RU HOROLOGII

$02^h\ 45^m\ 00^s\ -63^\circ\ 48'.6$ (1950.0) $14^m.0$–$16^m.9$ (photographic)

RU Horologii is one of the fainter southern U Geminorum variables and was found on a series of photographic plates taken by Hughes-Boyce[49] in 1942–43. As will be seen, it has a small photographic amplitude of 2.9 magnitudes from which we may presume a somewhat smaller visual range if the spectrum is similar to those of the other dwarf novae which have been examined spectroscopically. It lies close to the fifth magnitude star ν Horologii. So far, it has not been extensively observed, presumably because of its faintness, and the mean period is unknown.

AG HYDRAE

$09^h\ 48^m\ 13^s\ -23^\circ\ 30'.9$ (1950.0) $14^m.3$–$(16^m.0$ (photographic)

This particular variable has been recognised as a U Geminorum star for more than thirty years, having been found as long ago as 1936 by Hughes-Boyce[50]. Again, however, we are faced with the problem of having few observations available for discussion, due mainly of course to its faintness. Since the number of large instruments available for variable star work is limited, and those in the professional observatories are engaged on other fields of stellar research, it seems unlikely that further details of this, and other faint members of the class, will be forthcoming in the near future.

CT HYDRAE

$08^h\ 48^m\ 27^s\ +03°\ 19'.0$ (1950.0) $14^m.5$–($16^m.5$ (photographic)

Once again, this is a variable about which we know little. Discovered by Hoffmeister[51], the few maxima which have been observed, all photographically, have been of the abrupt type with evidently a steep rise to maximum, behaviour which is clearly typical of the dwarf novae. Very large apertures are again necessary for satisfactory observation of this star.

EX HYDRAE

$12^h\ 49^m\ 42^s\ -28°\ 59'.3$ (1950.0) $11^m.5$–$13^m.3$ (photographic)

EX Hydrae belongs to the fairly small group of U Geminorum variables possessing very small amplitudes. In this respect, it closely resembles V591 Centauri which was discovered at the same time as EX Hydrae by Huruhata[27]. If it obeys the cycle-amplitude relationship, even approximately, we would expect it to have a mean cycle length which is quite short since the visual amplitude is likely to be even smaller than the 1.8 magnitudes photographic given above. At present, however, there is still not enough observational data to confirm this. The variable lies a little too far south to be readily observable from northern European latitudes and being not too far removed from the ecliptic is totally unobservable during the autumn months. It is, however, now being actively observed by the Variable Star Section of the Royal Astronomical Society of New Zealand and a reasonably complete light curve should soon be available.

VW HYDRI

$04^h\ 09^m\ 29^s\ -71°\ 25'.9$ (1950.0) $8^m.4$–$13^m.9$ (visual)

The second brightest of all the known dwarf novae when at maximum, this variable has been closely studied by observers in the southern hemisphere, in particular by the Variable Star Section of the Royal Astronomical Society of New

Zealand[23]. The discovery of this U Geminorum variable, by Hughes-Boyce[49], was comparatively recent which is perhaps a little surprising in view of its general brightness and it is consequently only within the last fifteen years or so that a reasonably complete picture of its light variations has been obtained, due mainly to the efforts of Jones[23]. The amplitude is similar to that of SS Cygni but the mean period of only 32.6 days is appreciably shorter and for this reason it does not fit well onto the cycle-amplitude curve given for these variables in Chapter Two.

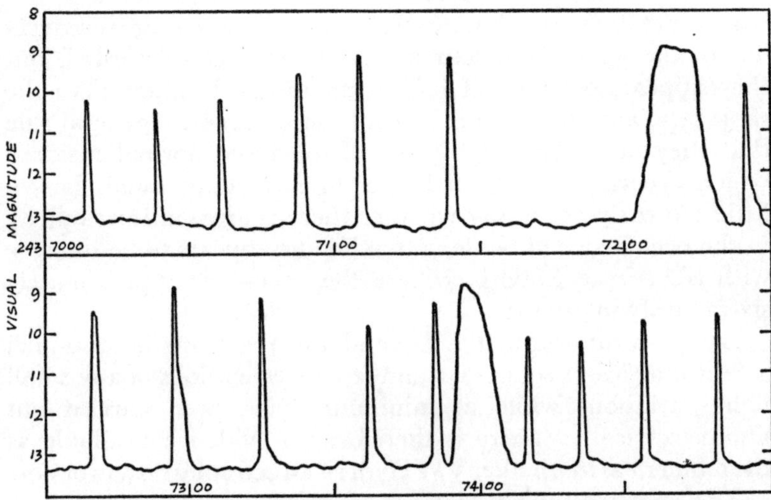

Fig. 42 – Light curve of VW Hydri (*Courtesy of the Variable Star Section of the Royal Astronomical Society of New Zealand*).

From the light curve (Fig. 42), it will readily be seen that two very distinct types of maxima are exhibited by this dwarf nova, these being quite dissimilar in shape, duration and maximum brightness. The long, fairly flat, maxima are of much less common occurrence than the short variety and it is perhaps inevitable that VW Hydri should be compared with SU Ursae Majoris although the latter star has a much shorter mean period of 16 days and is some three magnitudes fainter at maximum.

The rise to these rarer maxima is normally slow (very like the anomalous maxima of SS Cygni), the variable often taking three or four days to reach 8.5 magnitude. After fluctuating by a few tenths of a magnitude for several days, a slower decline sets in. The more common type of maximum is characterised by a very steep rise to a sharp peak and a somewhat slower fall with the star attaining a brightness of only about 9.1 magnitude, remaining at this peak for only a day or so before fading.

Although the variable has been comprehensively observed for only a comparatively short period compared with such dwarf novae as SS Cygni and U Geminorum, it is possible to indicate certain features of the light variations of this star which are worthy of note. It is quite true that relatively few of the long, flat maxima have so far been recorded, but they appear to be about half a magnitude brighter than the short type and in addition, it would seem almost a general rule that they are preceded by two of the more normal maxima which are separated by a longer interval than usual. It is a little too early yet to say whether there is any semi-periodicity in the occurrence of the long maxima as appears to be the case with SU Ursae Majoris (where they occur at approximately six-monthly intervals).

Lying so far south it is beyond the reach of the 200-inch reflector at Mount Palomar and few investigations of any small light variations while at minimum have been carried out photoelectrically. There is therefore no evidence available at the moment as to whether VW Hydri is an eclipsing system or not.

The spectroscopic variations of this star seem to be very similar to those of SS Cygni and U Geminorum. At, or near, maximum light the spectrum has a continuous distribution corresponding to that of a type B star, crossed by broad, shallow lines of hydrogen in absorption. At minimum there are equally broad, bright Balmer lines of hydrogen, weaker emission lines of neutral helium and also the bright H and K lines of singly ionised calcium. As for most of the dwarf novae, the apparently continuous background, at minimum, has an intensity distribution very like that of a late G-type dwarf. From this evidence, we are justified in assuming that the bright component is an sdBe star like those of the other U Geminorum variables; the precise spectral and luminosity class of the secondary has, however, yet to be determined.

The field of this variable is fairly open and there are several faint stars which may be used as comparisons when the star is at minimum.

AY LACERTAE

$22^{h}\ 20^{m}\ 22^{s}$ $+50°\ 08'.3$ (1950.0) $15^{m}.0$–($17^{m}.5$ (photographic)

A probable member of the U Geminorum class, AY Lacertae has been described by Himpel[52] but owing to its extreme faintness at all phases of its light variation, so few observations have been obtained that it is still impossible to assign it unambiguously to the dwarf nova class. Further investigation with large instruments is required before any definite conclusions can be reached about this star. It may possibly be a recurrent nova or one of the short-period irregular variables similar to RW Aurigae. AY Lacertae lies within the Milky Way, in a very rich field of faint stars and consequently presents some trouble in the way of identification, even when at maximum. Evidently it is not a variable which lends itself to visual observation and it is accordingly very doubtful if a complete light curve, from which its true nature may be established, will become available in the near future.

EG LACERTAE

$22^{h}\ 48^{m}\ 31^{s}$ $+54°\ 54'.8$ (1950.0) $15^{m}.5$–($17^{m}.5$ (photographic)

EG Lacertae is yet another of the large number of variable stars of all types to be discovered by Hoffmeister[21]. It is even fainter at maximum than the previously mentioned star and no details are yet available concerning its light curve. In spite of these difficulties, the series of plates on which it appears is sufficiently complete for it to be included with a fair degree of confidence among the U Geminorum variables. The rise is very rapid and the period observed between successive appearances short enough to rule out the possibility of it being a recurrent nova. Even when at maximum, the variable is at the very limit of spectroscopic detection with slit spectrographs.

T LEONIS

$11^h\ 35^m\ 53^s\ +03°\ 38'.9$ (1950.0) $10^m.1–15^m.4$ (photographic)

We now come to a very peculiar variable whose nature has been debated for many decades. Very few maxima of this star have ever been observed and for many years it was considered to be constant in brightness at 15.4 magnitude although minor fluctuations have been recorded at times. As long ago as 1865, however, it was reported as bright as 10.1 magnitude by Peters[53] but any similarity to U Geminorum, discovered only ten years earlier, was not recognised. Looking back on subsequent events, this is not really surprising. U Geminorum was first thought to be a fairly typical nova and it was not until its reappearance three months after discovery that this idea was shown to be untenable. In the case of T Leonis, the star faded quite rapidly after attaining its maximum brilliance and was not seen again (behaviour which is typical of novae). There was therefore no reason to suspect that it was anything else but a nova, several of which were known at the time of Peters' observations. Even as recently as twenty years ago, it was tentatively classified as being constant[54].

Now if we accept the original observation made by Peters and take this in conjunction with the fact that no subsequent maxima have been positively identified, we would not be justified in placing T Leonis in the dwarf nova category for in the absence of any further confirmatory evidence, all of the observational data can be explained on the basis of a faint nova. On what then do we base its inclusion among the dwarf nova? The answer lies in the spectrum of this star. If we examine high-dispersion spectrograms of the variable at minimum (we have, of course, no spectra of it when at maximum), we find a very striking resemblance to those of other U Geminorum variables.

It has been suggested by Brun and Petit[55,56] and Kraft[17] that a condition for membership of the dwarf nova class is that the spectrum at minimum should be of a particular kind, namely one showing emission lines of the type exhibited by such stars as U Geminorum (although not necessarily double as in this case) and other undisputed members of the class. T Leonis does indeed show these features in its spectrum and on this basis, it is now generally accepted that it belongs to this group and not to the novae.

High-dispersion spectrograms at minimum have been recently obtained by Kraft[17] using the 100-inch reflector at Mount Wilson with the Newtonian focus spectograph. The Balmer jump in emission is faintly visible in the spectrum but owing to the convergence of the lines before the limit of the series is reached, the Balmer lines blend into the background continuum.

The star lies almost exactly on the ecliptic, close to the sixth magnitude star 89 Leonis in a fairly open field. Being a zodiacal star it is unobservable from about the beginning of June to the middle of August. Large apertures are necessary to observe T Leonis at its present magnitude, but should it ever experience any further outbursts similar to that observed in 1865, it will be well within the reach of small telescopes and easily identified owing to the paucity of bright stars in the field. If we are to obtain any further worthwhile data on this very atypical U Geminorum star it is essential that observation of it should not be neglected. It is quite probable that some maxima have been completely missed for this very reason.

X LEONIS

$09^h\ 48^m\ 20^s$ $+12°\ 06'.7$ (1950.0) $12^m.0$–$15^m.4$ (visual)

One of the earliest U Geminorum variables to be discovered, by Metcalf[57] in 1907, this star has been continuously observed for many years, particularly by the Variable Star Section of the British Astronomical Association and the American Association of Variable Star Observers in the northern hemisphere and the Variable Star Section of the Royal Astronomical Society of New Zealand in the southern hemisphere. Like U Geminorum and T Leonis, it is situated close to the ecliptic and is unobservable for part of each year, from the beginning of June to the middle of August. Since its period is far shorter than either of the above two stars, quite a high proportion of the maxima have been unavoidably missed. There is also a 6.6 magnitude star in the same high-power field as the variable together with three much fainter companions, all of which add to the difficulty of making accurate visual estimates of its brightness, especially when around minimum. It is often advantageous to use an occulting bar to eliminate the glare of the bright companion when observing X Leonis.

Rises of this star are quite frequent, the mean period being only 22 days. There are, of course, some deviations from this mean figure but these are usually quite small, amounting to only one or two days (a feature we might expect from a variable with a short mean period). On one or two occasions, however, these deviations have become so marked that they give the appearance in the light curve (Fig. 43) of maxima which occur in pairs, one short and one long. Such anomalous behaviour lasts for only a short while before the more regular outbursts begin once again.

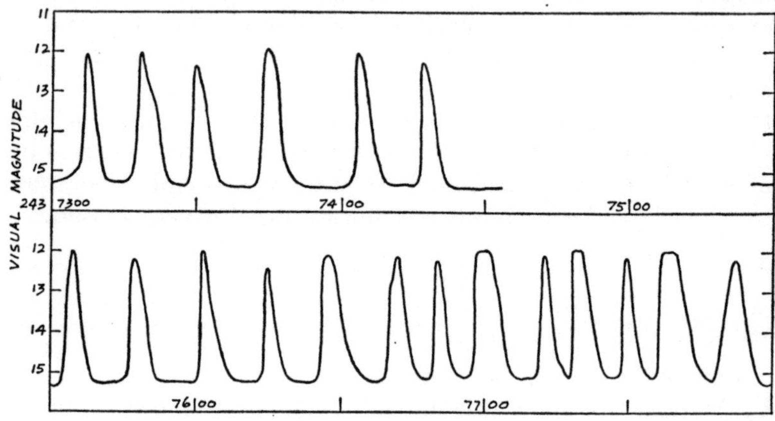

Fig. 43 – Light curve of X Leonis (*Courtesy of the British Astronomical Association*).

As with many of the U Geminorum variables, X Leonis shows a similar tendency for long and short maxima to alternate and in general the maxima, whether long or short, are reasonably flat. The minimum is more difficult to observe and as has been mentioned, observations at this phase are complicated by the presence of three faint stars situated close by. Although minor fluctuations have been recorded, these are not as pronounced as in the case of SS Cygni, for example.

The spectrum of X Leonis at maximum has been described as being essentially continuous by several observers. Elvey and

Babcock[8] in the course of their monumental survey of the dwarf novae and related objects, observed Hγ, Hδ and Hε as definite absorption lines when X Leonis was at maximum, but there was little, or no, indication of the presence of either Hα or Hβ. The lines are wide, diffuse and quite faint and from the distribution of energy in the underlying continuum, the spectrum has been estimated as like that of a late A or early F-type star, considerably redder than either U Geminorum or SS Cygni at maximum. Herbig[16] too, has recorded the spectrum of this star when at minimum using a nebular spectrograph with the 80-inch Crossley reflector. No emission lines have been found, the continuum being somewhat bluer than is usual for this class of variable.

TU LEONIS

$07^h\ 27^m\ 00^s$ $+21^\circ\ 36'.8$ (1950.0) $11^m.7$–$14^m.9$ (photographic)

TU Leonis was tentatively assigned to the U Geminorum class of the dwarf novae by Parenago[58] in 1934 but in spite of its relative brightness, few details of its mean period and spectrum have been recorded in the literature. Being another zodiacal variable, like X Leonis, it is impossible to observe this star for about three months in the year during the late spring and early summer.

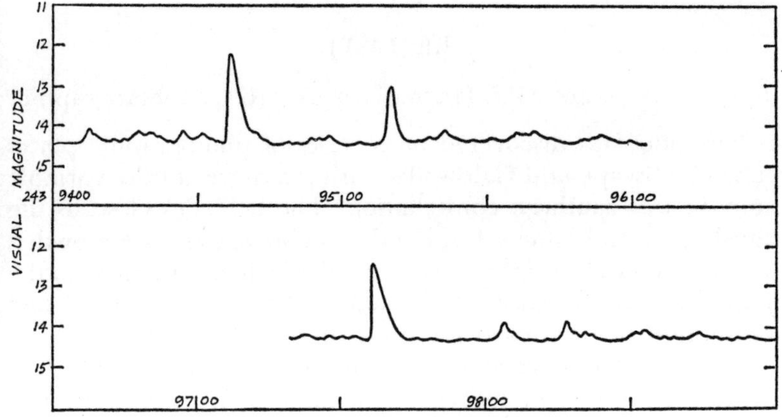

Fig. 44 – Light curve of TU Leonis.

The light curve given in Fig. 44 has been drawn from 375 visual observations made by the author during the period 1967–68. It is clear that the maxima are of the U Geminorum type, characterised by steep rises and somewhat slower declines. The star remains at maximum for only a short time and the three outbursts observed during this investigation are all of widely different magnitudes. Very marked fluctuations are evident during minimum, those observed at J.D. 2439413, 2439807 and 2439829 having been well observed with amplitudes of approximately 0.5 magnitude. From the distribution of observations, it appears unlikely that other major outbursts were missed during the period covered by the light curve, a fact which would suggest that although maxima of this variable are of fairly frequent occurrence, there are very wide deviations from any mean cycle. Taking all of the available evidence into consideration, the balance of evidence would seem to be in favour of TU Leonis being a member of the U Geminorum class and clearly this is a variable which will repay more extensive observation.

The field of the variable is readily located from NGC 2903 (a moderately bright spiral nebula) and the star itself lies close to two twelfth magnitude stars with which it may be compared when at maximum. There is also a 14.4 magnitude companion north following TU Leonis which is useful for comparison during the minimum phase.

BR LUPI

$15^h\ 32^m\ 31^s\ -40°\ 24'.6$ (1950.0) $13^m.1$–($16^m.4$ (photographic)

This star was discovered on a series of photographic plates taken by Swope and Caldwell[59] during a survey of the variable stars in this southern constellation. The field lies close to the third magnitude star γ Lupi and is unfortunately one crowded with moderately bright stars, several of which are close to the position of the variable itself, making both identification and estimation of brightness far from simple.

At present, the variable has not been followed through a sufficient number of successive maxima for a mean period to be determined.

SU LYRAE

$18^h\ 51^m\ 53^s$ $+36°\ 26'.6$ (1950.0) $12^m.0$–($16^m.5$ (photographic)

The light variations of SU Lyrae have been described by Chernova[60] but from the relatively small amount of data at present available for discussion, it is clear that this particular variable cannot be definitely assigned to the U Geminorum class. It lies in the same field as the bright double star δ^1 and δ^2 Lyrae and is surrounded by a host of faint stars. Once again, we are faced with the problem of having too few observations at maximum for any mean period to be derived. It is quite possible that the star may be a short period variable, for example a Cepheid or one of the various irregular types and until further observational data or spectroscopic information is forthcoming, this point is unlikely to be resolved. Unfortunately, SU Lyrae is too faint at minimum for high-dispersion spectrograms to be obtained with existing equipment (as we have seen earlier, an emission-type spectrum of the star at minimum is one criterion for its inclusion in the dwarf nova class). In all probability, we shall be forced to rely upon more extensive visual observations of this star to provide the answer.

AY LYRAE

$18^h\ 42^m\ 43^s$ $+37°\ 57'.1$ (1950.0) $12^m.6$–$16^m.1$ (photographic)

A study of the light curve of AY Lyrae (Fig. 45) reveals that this variable is a typical U Geminorum star. The minimum is very faint, the visual magnitude at this phase being about 15.6 and accordingly the star has not been as comprehensively observed at minimum as would be wished although most of the maxima have been well followed during recent years, particularly by the American Association of Variable Star Observers[7]. Two distinct types of maxima have been recorded with the short, sharply-peaked variety predominating. The long, brighter kind (which are sometimes almost a magnitude brighter), are of much less frequent occurrence, the star being much like VW Hydri and SU Ursae Majoris in this respect.

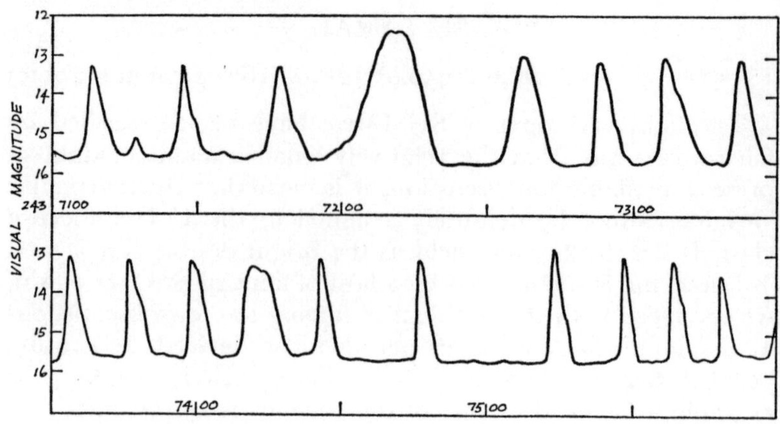

Fig. 45 – Light curve of AY Lyrae (*Courtesy of the American Association of Variable Star Observers*).

Like the preceding variable, AY Lyrae lies quite close to a bright double star, in this instance ζ Lyrae, the components of which are of 4.20 and 5.15 magnitude, their separation being 43·7 seconds of arc. The light curve of AY Lyrae shows several notable features. The long maxima, particularly the very bright ones, show a comparatively slow rise to peak brightness, rather like the anomalous kind recorded for SS Cygni. The minimum is generally fairly flat although one or two flare-like eruptions have been observed and the time spent at minimum varies quite appreciably from one cycle to the next. In one or two cases, a series of short maxima, occurring at intervals somewhat shorter than the mean period of 23 days, have been recorded in which there is a progressive diminution in the maximum brightness. Such a series is clearly shown in the accompanying light curve between J.D. 2437525 and J.D. 2437585.

Elvey and Babcock[8] found that the spectrum at minimum, unlike that of the other dwarf novae so far investigated, contains no bright emission lines but instead shows an absorption spectrum which may be approximately described as type G. It is quite possible that the blue component is relatively fainter than the G-type companion, maybe by as much as two magnitudes. If this is indeed the case, then the true visual range is greater than five magnitudes.

A more recent investigation of the spectrum of this star by Kraft[17] has not confirmed the absence of emission lines at minimum but we must bear in mind that AY Lyrae is extremely faint at this phase and lies in a very crowded field and it is possible that the star has been misidentified by the earlier investigators. Even if the original result should be confirmed and the presence of an emission-type spectrum at minimum does not hold for this particular variable, there is no doubt whatsoever from the light curve that it is a typical U Geminorum star.

CY LYRAE

$18^h\ 50^m\ 39^s\ +28°\ 41'.4$ (1950.0) $13^m.3$–$16^m.8$ (photographic)

This particular variable is one of the shorter-period U Geminorum stars with a mean period of only 17 days, very similar to that of SU Ursae Majoris. There are several bright stars in this field, together with a score or more which are between 12.0 and 14.5 magnitude. Even with large instruments and high magnifications, this makes visual identification and estimation of brightness around minimum troublesome.

The light curve (Fig. 46) shows the typical maxima of this variable, from which it will be seen that their shape varies widely. It is not easy to differentiate between long and short maxima and in almost every case there are pronounced

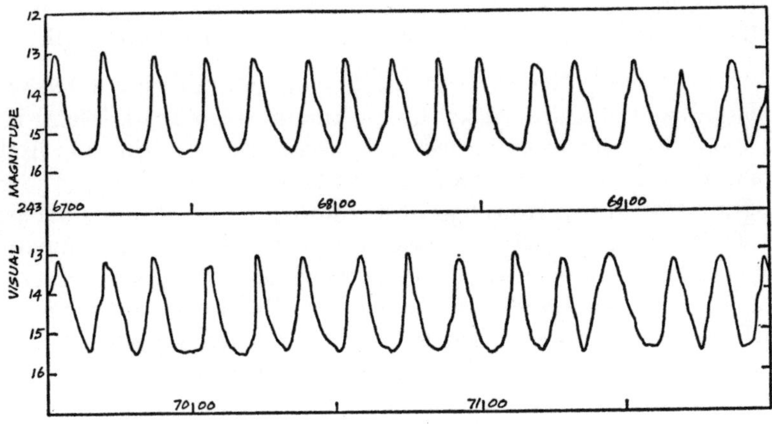

Fig. 46 – Light curve of CY Lyrae.

irregularities on both ascending and descending branches. The stay at minimum is very short, often only one or two days.

The spectrum of this star at minimum, as observed by Elvey and Babcock[8], is either continuous or shows broad, shallow Balmer lines. From the intensity distribution in the continuum, it has been estimated that the spectrum corresponds to that of an A-type star.

DM LYRAE

$18^h\ 56^m\ 49^s$ $+30°\ 11'.4$ (1950.0) $13^m.7$–$17^m.1$ (photographic)

DM Lyrae was found by Ahnert[47] during a very extensive survey of this small constellation. From the photographic estimates made, it is possible to construct a partial light curve and on the basis of the results obtained, it is possible to assign this variable to the U Geminorum class with a reasonable degree of certainty. Unfortunately, there is insufficient data from which a mean period may be derived.

The spectrum of DM Lyrae has not been fully investigated and at minimum is probably only within reach of slitless spectrographs using large apertures and long exposures. The field of this variable may be found from λ Lyrae, lying just to the south of this red star.

LL LYRAE

$18^h\ 33^m\ 29^s$ $+38°\ 17'.3$ (1950.0) $12^m.8$–$17^m.1$ (photographic)

This variable lies close to the first magnitude star α Lyrae (Vega) and was discovered by Hoffmeister[48] in 1950. There is a close thirteenth magnitude companion, necessitating the use of fairly high powers to separate the two stars. A portion of the light curve drawn from 658 observations by the author during 1965–68 using a 13-inch reflector is shown in Fig. 47. The mean cycle averaged over the four years is one of approximately 60 days but since the variable was invisible in this instrument when at minimum (indicated by the dotted regions of the curve) and coverage of the star was by no means as complete as is desirable for a variable of the cataclysmic type, such a mean period must be accepted with some reservation.

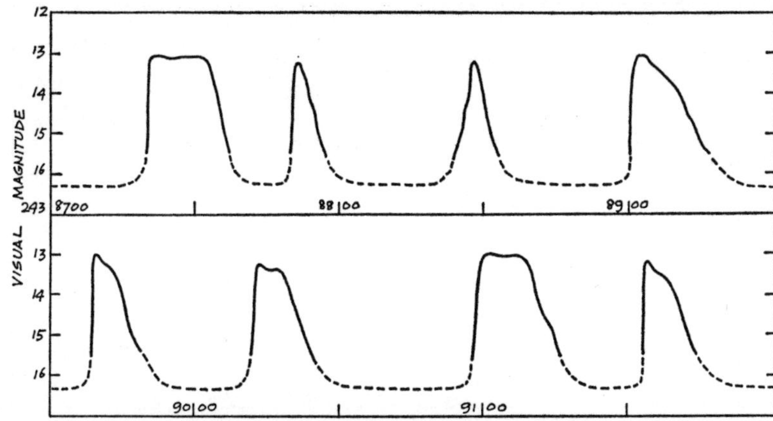

Fig. 47 – Light curve of LL Lyrae.

With one exception, that of J.D. 2438845, all of the maxima so far observed have been of the normal, abrupt type, typified by an extremely steep rise from minimum. The shape of the maxima vary enormously as will readily be seen from Fig. 47. The longer outbursts tend to be flat at maximum with generally a reasonably smooth decline, the shortest being sharply-peaked with again a fairly smooth fading to minimum. An intermediate type has also been recorded, however, in which there is a temporary pause on the descending branch about 0.3 magnitude below peak brightness.

Owing to the faintness of this particular variable when at minimum, very little is known of its behaviour during this phase. Apertures of at least 20 inches are necessary to observe it satisfactorily during these periods of its light variation.

CW MONOCEROTIS

06^{h} 34^{m} 22^{s} +00° 04′.7 (1950.0) 12^{m}.5–16^{m}.0 (photographic)

It was during the course of another of his photographic surveys that Ahnert[61] discovered CW Monocerotis which, from the light curve (Fig. 48), certainly appears to be of the U Geminorum type.

Of the two types of maxima observed, which may be somewhat loosely described as long and short, the latter are very definitely of the usual kind with a rapid rise to maximum and a

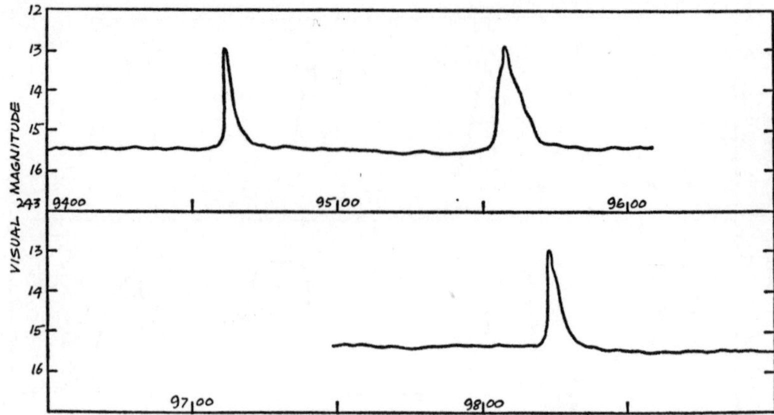

Fig. 48 – Light curve of CW Monocerotis.

somewhat slower decline. At minimum, the variable appears to fluctuate slightly around 15.6 magnitude, although it has been seen as faint as 16.1 magnitude by Cragg[7]. Unfortunately, the light curve is still a little too fragmentary for an accurate mean period to be derived.

CW Monocerotis lies in the galactic plane and is unobservable during part of the summer months. Two very close companions prove troublesome when the variable is around minimum.

EQ MONOCEROTIS

$06^h\ 55^m\ 14^s$ $-09°\ 43'.9$ (1950.0) $13^m.4$–$16^m.2$ (photographic)

The discovery of EQ Monocerotis is again attributable to Ahnert[62]. It has the very short mean period of only 13.9 days, only V485 Centauri being shorter than this among the U Geminorum stars with a period of approximately 10 days. In view of the shortness of the mean cycle, which incidentally agrees well with the cycle-amplitude relationship, it is probably not surprising that we find the maxima to be virtually all of the rapid type with very steep rises from minimum, this being generally the case with such members of the dwarf novae.

The star normally remains at maximum (Fig. 49) for only a day or so before beginning a fairly rapid decline. The field is very rich in stars and there are also three close companions which makes identification at minimum very difficult. As with the previous variable, EQ Monocerotis is unobservable for about two months during the summer.

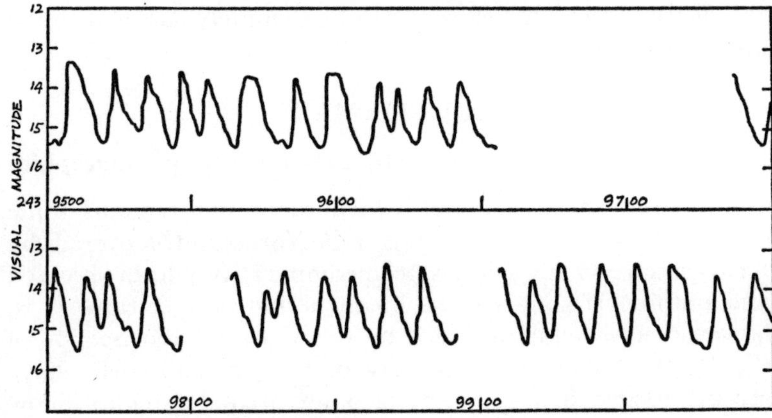

Fig. 49 – Light curve of EQ Monocerotis.

Because of the very short mean period of this star, we must consider the question of whether it is a true U Geminorum variable, or a member of the Z Camelopardalis group which, on average, have shorter periods than the main class of dwarf novae. Certainly the evidence we have at present would be against it being a Z Camelopardalis variable since no definite 'standstills' have ever been recorded. We must not forget, however, that (a) the star has not been extensively observed for any great length of time, (b) any 'standstill' will, if it follows the same general pattern as those of the majority of the Z Camelopardalis stars, occur below fourteenth magnitude and (c) the frequency of 'standstills' varies greatly from one variable to another. Clearly, a large number of additional observations are necessary before this question can be satisfactorily answered.

AB NORMAE

$15^h\ 45^m\ 48^s$ $-42°\ 56'.1$ (1950.0) $13^m.9$–(16.0 (photographic)

AB Normae was discovered by Swope and Caldwell[59] during the same photographic survey which resulted in the finding of BR Lupi. Unlike the latter star, the evidence for including AB Normae in the U Geminorum class is not at present conclusive and the possibility still remains that it may be merely an irregular variable of short period. Further data are clearly necessary before it can be definitely assigned to the present class.

HP NORMAE

$16^{h}\ 16^{m}\ 51^{s}$ $-54^{\circ}\ 45'.3$ (1950.0) $12^{m}.9$–$15^{m}.3$ (photographic)

A further variable about which there still remains some doubt as to its real nature is HP Normae. Discovered by Kruytbosch[63] in 1935, it has been comparatively little observed and nothing is known of its mean period or spectrum. It is, however, now being observed by the Variable Star Section of the Royal Astronomical Society of New Zealand who have recently begun a programme of observational work to follow all of the southern dwarf novae with maxima brighter than thirteenth magnitude and doubtless a reasonably complete light curve will be obtained in the near future when the type of light variation will be determined with certainty. The variable lies within the boundaries of the Milky Way, in a crowded field, and is consequently not an easy star to locate.

IK NORMAE

$16^{h}\ 21^{m}\ 27^{s}$ $-55^{\circ}\ 13'.4$ (1950.0) $13^{m}.0$–($15^{m}.5$ (photographic)

Discovered in 1942 by Hertzsprung[64], this U Geminorum variable lies in the same field as the preceding star but in spite of a similar scarcity of observations, it does appear from the photographs which have been taken that it is a dwarf nova. At present, the mean period has not been established nor have any details of its spectrum been published. The field is even more densely populated than that of HP Normae and two faint companions make identification hazardous at minimum.

TU OPHIUCHI

$16^{h}\ 23^{m}\ 41^{s}$ $-22^{\circ}\ 12'.6$ (1950.0) $13^{m}.6$–$16^{m}.5$ (photographic)

The light variations of this star, as far as they are known, have been described by Himpel[65]. It has been seen on only a few, isolated occasions and then only when around maximum. When at minimum, it is at the very limit of detection on the available photographic plates. The field may be readily located from the fourth magnitude star ω Ophiuchi but is rather low, even when on the meridian, for observers in northern European latitudes. Owing to its general faintness, no detailed spectroscopic analyses have been made of this variable.

V699 OPHIUCHI

$16^h\ 22^m\ 36^s$ $-04°\ 33'.8$ (1950.0) $13^m.8$–($16^m.0$ (photographic)

V699 Ophiuchi was discovered by Hughes-Boyce[66] in 1942 but unfortunately his observational data, although sufficient to establish quite definitely that it is a true U Geminorum variable, is not detailed enough to enable us to determine a mean period. During the winter months, the variable is observable only in the early morning but being well removed from the Milky Way, the field is reasonably open and easily found from δ and ε Ophiuchi.

V810 OPHIUCHI

$17^h\ 38^m\ 23^s$ $+07°\ 03'.6$ (1950.0) $14^m.8$–$15^m.9$ (photographic)

This extremely faint variable was found on plates taken by Hughes-Boyce and Huruhata[67] in 1941–42. Although assigned by them to the U Geminorum class, the photographic amplitude appears to be abnormally small for this type of star and from what we know of the spectra of the dwarf novae in general and their colour indices, we might expect the visual range to be even smaller, possibly no greater than 0.7 magnitude. Far more observational work is clearly required before its membership of the class is proved beyond doubt. If it should be shown to be a true dwarf nova, the amplitude is certainly one of the smallest for this class of variable. Unfortunately, it is beyond the reach of most amateur variable star observers and it is unlikely that this information will be forthcoming in the near future. The possibility mentioned earlier in the case of EQ Monocerotis is probably equally applicable here although for a different reason. Here, the very small amplitude may be indicative of a Z Camelopardalis variable but owing to the extreme faintness of this star, the problem will be correspondingly more difficult to solve.

CZ ORIONIS

$06^{h}\ 13^{m}\ 50^{s}$ $+15^{\circ}\ 25'.5$ (1950.0) $12^{m}.1$–$15^{m}.7$ (visual)

An excellent account of the light variations of this variable has been given by Rosino[68]. The mean period of 38 days fits well with the visual amplitude of 3.6 magnitudes and although there are naturally some annual deviations from this figure they are usually of a minor nature. During the summer months, CZ Orionis is not an easy star to observe from northern latitudes and consequently there are unavoidable gaps in the light curve.

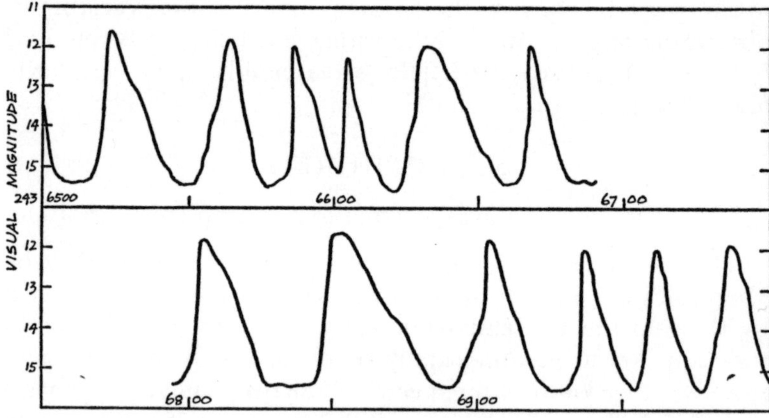

Fig. 50 – Light curve of CZ Orionis (*Courtesy of the British Astronomical Association*).

Nevertheless, since this star has been reasonably well observed for a number of years, initially by the American Association of Variable Star Observers and the Variable Star Section of the Royal Astronomical Society of New Zealand, and more recently by the Variable Star Section of the British Astronomical Association, we now know quite a lot about its general behaviour.

The maxima, like those of the majority of the U Geminorum

variables, of two kinds – long and short – characterised almost invariably by a steep rise and a slower decline to minimum. The light curve given in Fig. 50 illustrates well the general characteristics of this star.

The field is readily found from 69 and 72 Orionis and there are several nearby comparison stars whose magnitudes have been accurately established which are useful for estimating the brightness of the variable when below maximum.

V344 ORIONIS

$06^h\ 12^m\ 28^s\ +15°\ 29'.3$ (1950.0) $14^m.0$–($16^m.0$ (photographic)

This particular variable has been observed only on rare occasions since it was found by Hoffmeister[21], the recorded maxima being so scattered that it is impossible at present to derive a mean cycle. Its rise from below the limiting magnitude of the plates to maximum brightness is evidently rapid, indicative of the nova-like outbursts of this class of star. Its position is very close to that of the preceding variable and since it lies in the galactic plane, there are numerous stars in the field with the result that positive identification is again a troublesome feature of its observation. No spectroscopic details are available for V344 Orionis even when at maximum.

V350 ORIONIS

$05^h\ 37^m\ 51^s\ -09°\ 44'.2$ (1950.0) $10^m.9$–$12^m.6$ (photographic)

There has been a great deal of uncertainty about the nature of this star since its discovery in 1929 by Hoffmeister[69]. Some of the early observers, in particular Tsesevitch[70] considered it to be either an irregular or long period variable. More recent observations by Soloviev[71] and Ashbrook[72] have resulted in its inclusion in the U Geminorum class.

The accompanying light curve, obtained by the author from 377 observations made during 1967–68 (Fig. 51), would indicate a fairly long mean period for this variable if only the major outbursts are used for its determination; if the very short, minor eruptions are taken into consideration, then the mean cycle is appreciably shorter. The minima are, as a rule, quite disturbed, with marked fluctuations of the order of

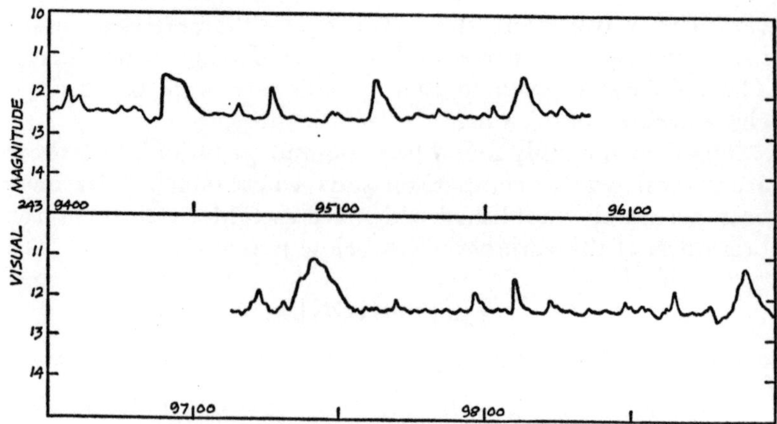

Fig. 51 – Light curve of V350 Orionis.

0.4 magnitude occurring at short, irregular intervals. In general, the major outbursts show a fairly rapid rise to maximum although the amplitude of little more than one magnitude is abnormally small for this class of variable.

The field lies approximately midway between Rigel and Kappa Orionis and there is a 6.36 magnitude star very close to the variable which hampers accurate visual estimation of its brightness unless an occulting bar is employed. The star is one of an irregular tetrahedron of faint stars and as the whole of the light cycle is well within the reach of moderate-sized instruments, V350 Orionis is particularly well adapted to visual observation. This star is certainly one which will repay continued study in order to firmly establish its true nature.

AS PAVONIS

$18^h\ 07^m\ 43^s$ $-59°\ 23'.0$ (1950.0) $14^m.0$–($16^m.5$ (photographic)

This southern variable is a probable member of the U Geminorum class but being so faint, it will require many more observations than are available at present to decide its type with any degree of certainty. Boyd[73] who discovered the star during a preliminary photographic survey of this region close to the southern Milky Way, first suggested its inclusion in the dwarf nova class on the basis of its light variations.

RU PEGASI

$22^h\ 11^m\ 36^s$ $+12°\ 27'.2$ (1950.0) $10^m.0$–$13^m.1$ (visual)

RU Pegasi has been known for more than sixty years and has also been continuously observed since its discovery. As a result we know almost as much about its light variations as we do of SS Cygni. Unlike this variable, however, RU Pegasi is not a circumpolar star and during the early summer months there is a short period when it is observable only just before dawn. As has been mentioned earlier in Part One, the mean period of 70 days appears too long for a visual amplitude of only 3.1 magnitudes – and only on rare occasions has it been seen as faint as 13.1 magnitude. More usually, this star varies between 10.3 and 12.6 magnitude and although a small telescope is sufficient to show it when at maximum, a moderately large instrument and high powers are necessary when the star is near minimum since there is a 12.5 magnitude companion extremely close to the variable and the two must be clearly separated for an accurate visual estimate of the brightness to be made at this phase. This may be a possible reason for the discrepancies which are sometimes found among the various estimates made around minimum. When at maximum, both of these stars may be estimated as a single object without any appreciable loss of accuracy.

Generally speaking, the variable is fairly regular in its outbursts and in spite of the rather long mean period, deviations from it are usually quite small although as is apparent from the light curve (Fig. 52), there are exceptions to this. One peculiar feature of the maxima of this star will be clear from an examination of the accompanying light curve. The maxima may be broadly divided into two types, those which are ostensibly flat and others which are more rounded at the peak but in every case the rise from minimum is quite slow, often with pronounced irregularities. It is doubtful if any of the characteristic steep rises have been recorded for RU Pegasi during all of the years it has been under observation.

The minima are reasonably flat but as we have already seen it may be that this apparent smoothing out of any fluctuations is due to the proximity of the 12.5 companion which makes accurate observation of this phase difficult.

The spectrum of RU Pegasi has been recorded by several observers. Elvey and Babcock[8] noted the presence of bright Balmer lines and neutral helium lines when at minimum, together with a faint K line of singly ionised calcium, so although there is some evidence that the blue star is relatively faint compared with the cooler secondary companion, the difference is therefore unlikely to be in excess of two magnitudes.

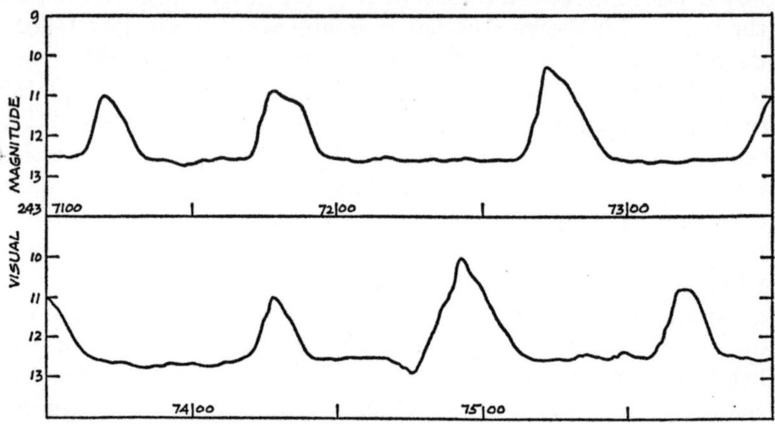

Fig. 52 – Light curve of RU Pegasi (*Courtesy of the British Astronomical Association*).

The true visual range is therefore slightly more than five magnitudes and from this observation, we see that the curious anomaly of the long mean period associated with a fairly small amplitude is explained. The minimum actually observed is chiefly due to the red component, the true minimum of the white dwarf companion being almost certainly nearer to magnitude 15.0. This explanation, of course, is based upon the view that the white dwarf star is the seat of the nova-like outbursts. In view of Krzeminski's observation that in U Geminorum it is the red component which is responsible for the eruptions, this view may have to be modified in the light of future work.

Three years before Elvey and Babcock carried out their spectroscopic survey of the dwarf novae, Joy[3] was able to show that the spectrum of this variable is composite, that at minimum it possesses the absorption lines of a star of spectral type dG3 together with the more usual emission features. Apart from SS and EY Cygni, this is the only U Geminorum variable in which the absorption spectrum of the late-type component appears in the photographic region.

The high-dispersion spectrograms obtained by Kraft[17] using the prime-focus spectrograph of the 200-inch Palomar reflector, have yielded a great deal of valuable information concerning this variable. The K line of ionised calcium, although normally single, appears double on certain of the spectrograms. The spectrum of the secondary component, too, has been found to vary between dG8 and dK0 depending upon the phase of the light variations. Not only this, but the luminosity class of this star varies quite appreciably throughout the light cycle from Class IIIn to Class Vn. The emission lines themselves have somewhat amorphous edges which makes accurate measurement of their widths extremely difficult and, as with other U Geminorum variables, the higher members of the Balmer series converge quite rapidly although it does seem from the observations which have been made that the Balmer jump is present in emission.

The radial velocity curves, both for the emission and the absorption lines are symmetrical and have been given earlier (Fig. 16). This symmetry indicates that here the orbit is essentially circular. As we have mentioned elsewhere, all of the U Geminorum variables for which radial velocity curves are available indicate a circular orbit, in contrast to that of RX Andromedae which seems to be pronouncedly eccentric.

From the fact that the luminosity class of this star can be measured quite accurately (it is actually determined from the relative strengths of the ionised strontium lines at λ4077 and λ4215), and this class varies with aspect, we have fairly strong evidence that within this particular system, the red secondary component completely fills its lobe of the equipotential surface. As Kraft has found, the highest luminosity corresponds to the phase when the two components are in conjunction with the cooler companion behind.

We have already seen in Chapter Four that the estimated minimum radius of the late-type component of RU Pegasi is 5.9×10^{10} centimetres while the calculated radius of the inner lobe of the equipotential surface is only 5.1×10^{10} centimetres.

Now it can be shown that at superior conjunction there is a reduction in the effective gravity at the surface of the secondary and we would, on the basis of this, expect a dwarf star which fills its lobe of the equipotential surface to exhibit a higher luminosity than normal. If this were a normal dwarf star with a spectral type of dG8 and a luminosity class on average of IVn, its radius would be much greater than that which is actually found, lying between 1.0 and 2.0×10^{11} centimetres and its mass would be nearer 12 times that of the sun. Such a result is immediately ruled out by the observational fact that the absolute magnitude at minimum is between +7.5 and +9.5. It would seem therefore, that this abnormally high luminosity is spurious, resulting from the lowering of the surface gravity of this star by the high centrifugal acceleration.

Because of its overall brightness and the fact that we are able to distinguish the absorption spectrum of the secondary star in the photographic region of the spectrum, we are able to make a direct estimate of the masses of the two stars in this system. Although there is, at present, no evidence of an eclipse (which would provide us with valuable additional information), the presently accepted view is that both components have masses somewhat less than that of the sun, between 0.6 and 0.9 solar masses with the secondary probably being slightly the more massive of the two. A rough estimate of the inclination of the orbit has been derived in Part One, this being approximately 60°, and although there are necessarily some uncertainties associated with this value, it lends support to the argument that this variable does not eclipse.

UV PERSEI

$02^h\ 06^m\ 44^s$ $+56°\ 57'.1$ (1950.0) $12^m.1$–$16^m.8$ (visual)

Like RU Pegasi, this variable has been known for more than half a century but is one of the more difficult of the dwarf novae to observe visually. Not only is the star so faint when at minimum, but the rises of UV Persei are so very infrequent that it is seldom seen at maximum. All of the maxima which have so far been recorded have been of the rapid type with the star remaining at maximum for only one or two days before fading, the descent to minimum being usually smooth and relatively slow compared with the rise (Fig. 53).

The mean cycle has been given as about 300 days but since many maxima have undoubtedly been missed due to adverse observing conditions, this figure will doubtless have to be revised, possibly quite appreciably, as more observations become available using large instruments. There is also a further problem associated with the determination of an accurate mean period. The extreme brightness of the star at maximum is 12.1 magnitude but unfortunately, the majority of the maxima have proved to be much fainter than this, rarely being above thirteenth magnitude and this, coupled with the comparatively short period spent above minimum, makes its observation difficult. On at least one occasion, however, a double maximum has been recorded.

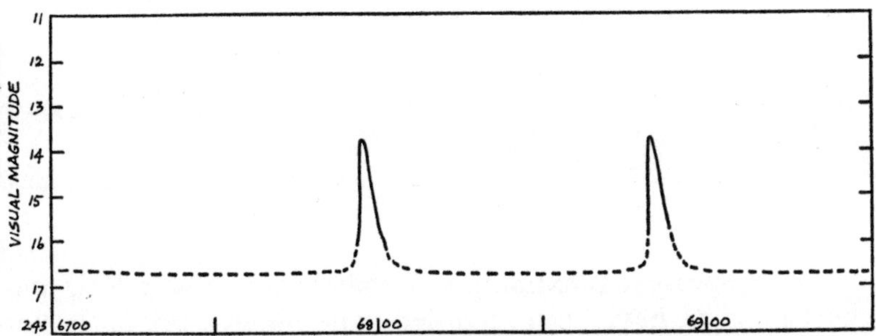

Fig. 53 – Light curve of UV Persei (*Courtesy of the British Astronomical Association and the American Association of Variable Star Observers*).

The field of UV Persei is most readily found from the bright double cluster *h* and χ Persei which are easily visible to the naked eye on a clear, moonless night, midway between Perseus and Cassiopeia. The variable itself lies almost midway between two eleventh magnitude comparison stars and there is a close companion of 15.0 magnitude with which it must not be confused when at minimum.

TY PISCIUM

$10^h\ 23^m\ 27^s$ $+32°\ 07'.5$ (1950.0) $12^m.5$–($15^m.0$ (photographic)

This particular U Geminorum variable has been described by Parenago[74]. It lies in a reasonably sparsely populated field and although there are four companion stars quite close to the variable and of about the same magnitude as TY Piscium when at a typical maximum, a fairly high magnification will separate them quite easily. Fig. 54 is a portion of the light curve of this star, drawn from 478 observations by the author during 1966–68 and would suggest a visual magnitude at minimum of about 15.3.

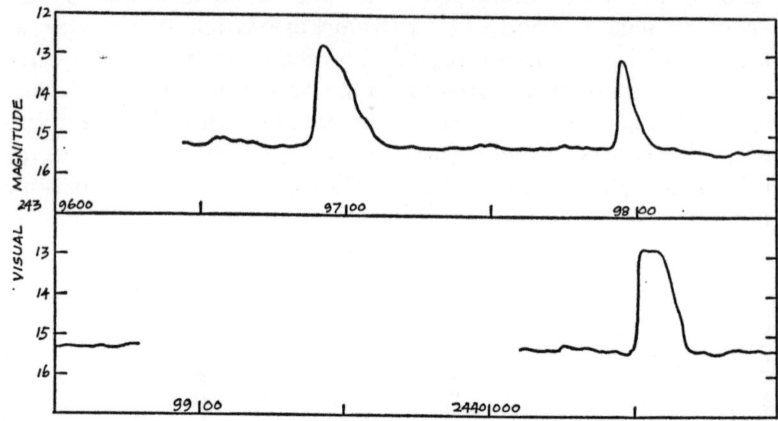

Fig. 54 – Light curve of TY Piscium.

The steep rise to maximum is readily seen and both long and short maxima have been recorded. The minima are fairly flat although small fluctuations have been noted, usually not amounting to more than 0.3 magnitude.

Unfortunately, this is another zodiacal dwarf nova and is totally unobservable for about three months of each year, from February to the middle of May. It is however, situated on the northern border of this constellation and consequently there are few observational difficulties associated with its position for observers in European latitudes.

BV PUPPIS

$07^{h}\ 46^{m}\ 58^{s}$ $-23°\ 26'.4$ (1950.0) $13^{m}.1$–$15^{m}.1$ (photographic)

Discovered by Dirks[75] in 1941, BV Puppis lies in a rather densely populated field, being a Milky Way star, but may be found from its proximity to the fairly bright nebula, M93.

The mean period given by Dirks, of 19 days is what we might expect for a U Geminorum star having such a small amplitude and in this respect it fits the cycle-amplitude relationship very well.

Apart from the original observations made at the time of its discovery, we know very little of its light curve although it is well within the reach of moderate-sized telescopes. The Variable Star Section of the Royal Astronomical Society of New Zealand, who are presently engaged on a programme of observational work to cover all of the dwarf novae in the southern hemisphere which are brighter than about thirteenth magnitude at maximum, will doubtless provide much needed data regarding this variable in the near future. From northern latitudes, it lies too close to the horizon, even when on the meridian, for satisfactory observation.

CL PUPPIS

$07^h\ 27^m\ 50^s\ -19°\ 21'.6$ (1950.0) $14^m.4$–($16^m.5$ (photographic)

This variable was found on a series of photographic plates taken by van Hoof[75] in 1940–41. Although the light variations would appear to be similar to those of a typical dwarf nova, the number of observations at present available for discussion are not sufficient to enable us to place it definitely in the U Geminorum class and many more are required before we can be absolutely certain of its type. Since we also know nothing of its spectrum, there remains the possibility that it may be a Cepheid having a short period or an irregular variable.

The star lies in the galactic plane, in the main stream of the Milky Way and is just within reach of observers in north European latitudes during the winter months.

RZ SAGITTAE

$20^h\ 01^m\ 02^s\ +16°\ 54'.3$ (1950.0) $12^m.2$–($16^m.0$ (photographic)

RZ Sagittae has been described by van Maanan[77] but although one of the brighter U Geminorum variables at maximum, we are not yet able to assign a mean period to its light variations. The variable has been observed for some years by the American Association of Variable Star Observers[7] and several maxima have been recorded, but the spread of observations, and their number, are not sufficient for a light curve to be constructed. The amplitude would appear to be at least four magnitudes.

The field may be located from the sixth magnitude stars 14 and 15 Sagittae and lies in the main stream of the Milky Way.

WZ SAGITTAE

$20^h\ 05^m\ 19^s$ $+17°\ 32'.8$ (1950.0) $7^m.0$–$16^m.1$ (photographic)

Strictly speaking, this variable should be described as a recurrent nova and not a member of the U Geminorum class at all. There are, however, certain peculiarities associated with this star which, as we shall see, justify it being considered in association with the dwarf novae. Two major outbursts of WZ Sagittae have been recorded, in 1913 and 1946, the visual amplitude being between seven and eight magnitudes which is a somewhat smaller range than for other recurrent novae. Since the star has been closely observed during the intervening period, it seems improbable that other maxima have gone unnoticed and consequently as far as the mean period is concerned (if we can use this term when there are only two maxima to be considered), the star would clearly appear to belong to the recurrent nova class.

Let us now examine the spectroscopic characteristics of this variable. At minimum, the absorption spectrum of a white dwarf star is clearly visible, together with double emission lines due to hydrogen, the latter characteristic of a system in which the plane of the orbit is inclined at about 90°. Greenstein[78] has also shown that the absolute magnitude at minimum must be quite faint, possibly even fainter than +10.0 which is more like that of a dwarf nova than a recurrent nova. Further confirmation of this comes from the work of van Maanan[79] who has determined a proper motion of 0.080 ± 0.004 second of arc. If the star had an absolute magnitude at minimum similar to those of the recurrent novae, between +6.0 and +7.0, the tangential velocity would be impossibly high for either a white dwarf star or a recurrent nova.

Using a special technique of trailing the spectrum across a

long slit, Kraft[80] has demonstrated the existence of an S-wave superimposed upon the double emission lines which has a period of 80 minutes. That WZ Sagittae is also an eclipsing binary has been shown by Krzeminski[81] who found a period of 81.5 minutes and a light curve during eclipse which, although of the W Ursae Majoris type, with continuous light variation, also exhibits very pronounced erratic photometric variations in the curve.

When we attempt to derive the masses of the two components, we come up against several problems which still await solution. No matter how the masses are calculated, that of the white dwarf companion always exceeds the theoretical upper limit of 1.2 solar masses. If we try to explain these higher masses on the assumption that the primary component is a neutron star as described by Burbidge[82] we again run into difficulties. In view of all this, it is not yet possible to put forward a satisfactory model for the WZ Sagittae system which explains all of the observational data.

Finally, one piece of evidence which suggests a very strong link with the dwarf novae is the absence of any lines due to ionised helium in the spectrum. These lines are clearly visible in the spectra of the recurrent and ordinary novae but not in the U Geminorum stars due, of course, to the higher surface temperatures prevailing in the primary components of the former stars.

WW SAGITTARII

$18^h\ 19^m\ 57^s\ -27°\ 27'.2$ (1950.0) $12^m.2$–($16^m.5$ (photographic)

This variable, discovered by Himpel[83] is another star whose inclusion in the dwarf nova class is still far from definite. It has been comparatively little observed since its discovery and it is not possible to draw a light curve from the number of observations which have been made, there being several large gaps, most of which are unavoidable owing to the proximity of the variable to the ecliptic. The number of successive maxima too, is insufficient for us to derive a mean period, if indeed it is a U Geminorum star. It lies in a very rich field of stars within the boundary of the Milky Way, in the direction of the galactic centre, close to the fourth magnitude star 24 Sagittarii.

V551 SAGITTARII

$17^h\ 57^m\ 35^s$ $-34°\ 35'.0$ (1950.0) $13^m.7$–($16^m.5$ (photographic)

Since its discovery by Swope[14] in 1936 during a comprehensive photographic survey of the variable stars in this large, zodiacal constellation, this variable has been observed on only a few isolated occasions and then only when at maximum. Accordingly, very few details of its light variations are known. Those rises which have been recorded seem to be of the abrupt, nova-like type characteristic of the U Geminorum stars and it is generally accepted that V551 Sagittarii belongs to this class.

V1089 SAGITTARII

$19^h\ 05^m\ 53^s$ $-17°\ 26'.8$ (1950.0) $13^m.8$–$16^m.6$ (photographic)

V1089 Sagittarii is yet another member of the U Geminorum class which have amplitudes of less than three magnitudes. This particular variable has unfortunately not been observed over a long enough period since its discovery by Uitterdijk[84] for its mean period to be determined. Unlike the other dwarf novae in this constellation, it lies on the border with Scutum, some distance from the Milky Way in a comparatively open field.

FQ SCORPII

$17^h\ 04^m\ 51^s$ $-32°\ 37'.8$ (1950.0) $12^m.0$–($16^m.5$ (photographic)

Observation of this U Geminorum variable, which was found by Swope[85], is seriously hampered by the presence of a bright companion which lies very close to the variable. Except with very high modifications, an occulting bar is of little use in this case. The mean period of only 24 days would suggest that the minimum photographic magnitude is not far below the limiting value given above and this is confirmed by the visual light curve (Fig. 55), where the magnitude at minimum would appear to lie between 15.3 and 15.6.

Most of the observed maxima are of the rapid type, remaining at peak brightness for only two or three days before a slower decline sets in. Two flat maxima have been definitely recorded, resembling the classical long kind of SS Cygni and U Geminorum. The decline is often far from smooth with a pronounced hump on the descent which may occur almost anywhere during the fall to minimum. In general, the derived light curve confirms the period of 24 days given by Swope.

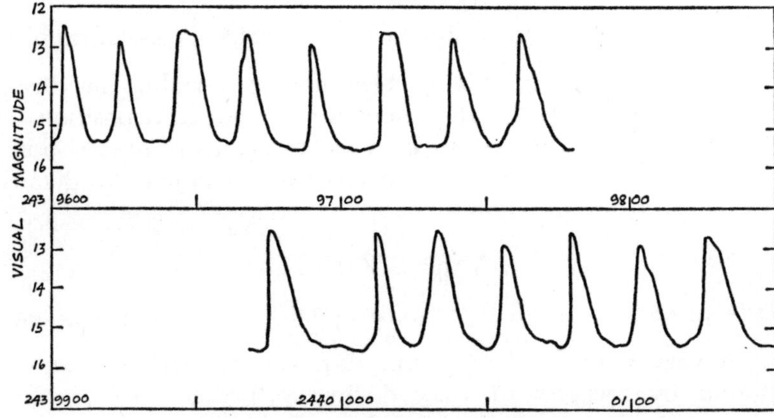

Fig. 55 – Light curve of FQ Scorpii.

Again, we are faced with all of the problems associated with a zodiacal variable lying near the galactic plane; a crowded field of faint stars and its inaccessibility during the winter months when the sun is passing through the constellation. Because of its southerly aspect it is also quite a difficult star for observers in the northern hemisphere. The field may be easily located from NGC 6558.

MM SCORPII

$17^h\ 27^m\ 12^s\ -42^\circ\ 09'.0$ (1950.0) $13^m.0$–($16^m.5$ (photographic)

MM Scorpii was found during an extensive survey of selected Milky Way fields in Scorpio carried out by Shapley and Swope[13]. Like the preceding variable, it lies in a densely populated field and again there is the added complication of a very close companion which makes observation of this star. particularly when near minimum, far from easy. As with most of the variable stars in this constellation, it is better suited to investigation by observers in the southern hemisphere and is almost certain to be included in the observational programme being initiated by the Variable Star Section of the Royal Astronomical Society of New Zealand mentioned earlier. The field is readily found from the second magnitude star θ Scorpii.

V478 SCORPII

$17^h\ 22^m\ 37^s\ -35°\ 29'.7$ (1950.0) $14^m.0$–($16^m.5$ (photographic)

The discovery of this U Geminorum variable has been reported by Swope[86]. At present we are not in a position to assign any mean period to it owing to a lack of observations covering a sufficient number of successive maxima. No details of its spectrum are known.

V598 SCORPII

$16^h\ 59^m\ 39^s\ -34°\ 36'.8$ (1950.0) $14^m.8$–($16^m.5$ (photographic)

A very faint U Geminorum star, this variable was again found by Swope[87] in 1943 during a further photographic survey of this constellation. It is beyond the reach of all but the largest instruments at minimum and being faint, even when at maximum, we know very little of its light variations and nothing of its spectrum. It lies just within the boundaries of the Milky Way and is not an easy variable to locate owing to the profusion of faint stars in this region.

V601 SCORPII

$17^h\ 00^m\ 41^s\ -36°\ 39'.4$ (1950.0) $15^m.0$–($17^m.0$ (photographic)

A second extremely faint variable found by Swope[87], this star has not yet been definitely assigned to the U Geminorum class although the rise to maximum does appear to be quite rapid, a characteristic indicative of these variables. It is quite possible, however, that as more observations become available, it may be placed in a different category altogether.

UZ SERPENTIS

$18^h\ 08^m\ 33^s\ -14°\ 56'.4$ (1950.0) $12^m.0$–$16^m.6$ (photographic)

UZ Serpentis is a fairly typical U Geminorum variable, the rises being on the whole quite steep with the star attaining maximum brightness within one or two days, the decline to minimum being markedly slower. A part of the light curve (Fig. 56), drawn from 497 observations by the author during 1965–68 illustrates the general behaviour of this star.

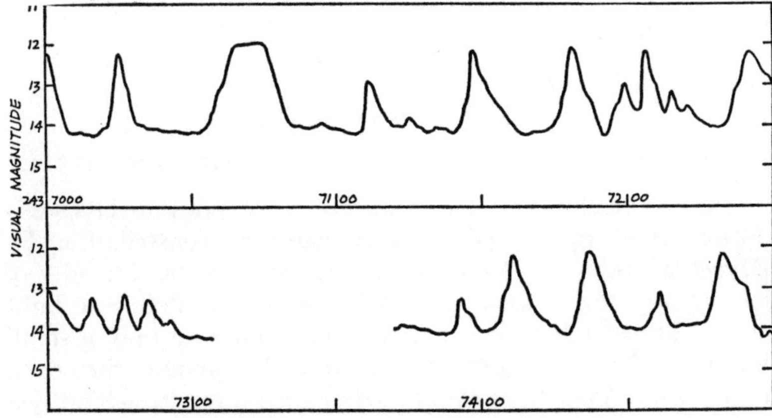

Fig. 56 – Light curve of UZ Serpentis.

The mean period averaged over the three years or so during which it was under observation is one of 43 days which is in good agreement with that of about 40 days given in the literature[54]. Although it is not a zodiacal star, UZ Serpentis is not readily observable during January and February, rising only an hour or so before sunrise but it is well placed for northern observers throughout the remainder of the year.

Herbig[88] has made a special study of the colour of this star, particularly for any interstellar reddening and concludes that the variable lies in front of a cloud of absorbing matter, the latter being not more than 650 light years distant. If this is the case, then we have an upper limit of its distance and we may obtain a reasonably accurate estimate of its absolute magnitude at minimum, the accepted figure being +10.1. As we have already seen, those U Geminorum variables for which a statistical parallax has been derived, have an average minimum absolute magnitude of about +7.5 to +9.5, so this independently derived value is in fairly good agreement with this, especially when we take into consideration the observed fact that there is almost certainly a spread of absolute magnitudes of these stars.

The field of UZ Serpentis is very crowded, a feature we would expect of a star which lies very close to the galactic plane and there are many stars close to the position of the variable of about the same magnitude as the star when around maximum.

WW TELESCOPII

$18^h\ 19^m\ 52^s$ $-55°\ 01'.2$ (1950.0) $14^m.6$–($17^m.5$ (photographic)

This particular variable is one of three possible U Geminorum stars discovered in this southern constellation by Boyce[89] in 1938–39. Again, we are faced with the difficulty of having too few observations with which to define a light curve. The star has been found at maximum on only a small number of photographic plates and the actual rise from minimum has not been clearly defined. As a result, it is not yet possible to place this variable with any degree of certainty in the dwarf nova category and far more observations using large instruments will be necessary before it can be firmly accepted as a U Geminorum star. The field may be found from the fourth magnitude stars θ Arae and ζ Telescopii.

YY TELESCOPII

$18^h\ 29^m\ 51^s$ $-54°\ 01'.5$ (1950.0) $14^m.4$–($17^m.5$ (photographic)

Although Boyce[89] has suggested that this variable may belong to the U Geminorum class, it now seems more likely that it was a faint nova as it does not appear to have been observed at maximum since its discovery thirty years ago. It is, of course, extremely faint at all phases of its light variation, the minimum being beyond the limiting magnitude of the series of plates taken at the time of its discovery. Consequently, much more data using the largest apertures available are desirable before any definite conclusions may be reached. Since it lies in the galactic plane within a few degrees of the Milky Way, its identity as a faint nova would seem to be a very definite possibility.

AK TELESCOPII

$18^h\ 39^m\ 47^s$ $-55°\ 03'.7$ (1950.0) $14^m.6$–$15^m.8$ (photographic)

AK Telescopii is the third possible U Geminorum variable discovered by Boyce[89] during his photographic survey of Telescopium. In this case, however, the available evidence is somewhat more in favour of it belonging to the dwarf nova class. The star is visible on most of the plates taken and the small amplitude of only 1.2 magnitudes clearly rules out the possibility of a nova. The number of observations made over the past thirty years, however, are not enough for a satisfactory light curve to be drawn and no mean period can be defined from those which are available.

VW TOUCANAE

$00^h\ 18^m\ 06^s$ $-74°\ 09'.6$ (1950.0) $15^m.4$–($17^m.5$ (photographic)

Found on photographic plates taken by Shapley and Hughes-Boyce[90] in 1933–34, the nature of this particular variable is still in doubt. Obviously, from its extreme faintness, it is not one which can be readily followed by visual observers and even when at maximum it can be observed only with very large instruments. It is quite possible that the star was a faint nova although it lies well away from the galactic plane, being only a few degrees from the Lesser Magellanic Cloud. The light variations would appear to rule out the possibility of it being a variable of the RR Lyrae type although it may be an irregular variable or even a very faint Cepheid.

SU URSAE MAJORIS

$08^h\ 08^m\ 05^s$ $+62°\ 45'.6$ (1950.0) $10^m.9$–$14^m.5$ (visual)

Discovered as long ago as 1908 by Madame Ceraski[91], this particular U Geminorum variable has been closely followed for more than half a century both by the Variable Star Section of the British Astronomical Association and the American Association of Variable Star Observers and as a result of the large number of observations recorded, the light curve is reasonably complete, especially over the past twenty years or so. The mean period of 16 days is very short for a variable of this type, however, and as the normal maxima are characterised by a steep rise, a stay of only one or two days at maximum and a fairly rapid decline, unfavourable observing conditions undoubtedly result in many maxima being missed altogether. In more recent years, with an increase in the number of observers using fairly large instruments, the position has improved considerably and there have been very few gaps indeed in the light curve (Fig. 57).

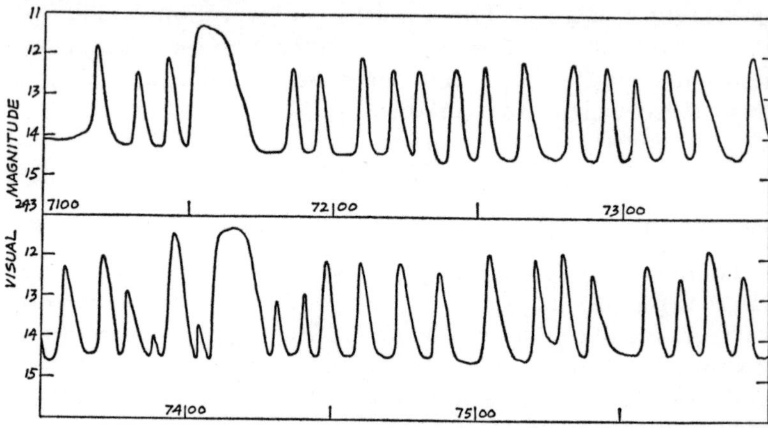

Fig. 57 – Light curve of SU Ursae Majoris (*Courtesy of the British Astronomical Association*).

Mention was made in Chapter Two of the two very distinct types of maxima exhibited by this variable; a very short, sharply-peaked kind which is by far the more common and a 'supermaximum' which occurs about every six months, during which the star remains above minimum for an average of 20 days or more and attains a brightness about a magnitude higher than at a normal maximum. From the data accumulated over the past thirty years, there would seem to be quite a marked regularity about the occurrence of these 'supermaxima' which has led to the suggestion that they may be at least semi-periodic and it is tempting to suppose that they form a kind of secondary period superimposed upon the primary mean period of 16 days. As yet, however, we have very little idea of the physical process operating within this system which may be responsible for these 'supermaxima'.

In the case of VW Hydri which exhibits a similar phenomenon we have seen that such anomalous maxima are usually preceded by three normal maxima more widely spaced than usual. So far, there is little evidence of this as far as SU Ursae Majoris is concerned although, as seen from the accompanying light curve, the light variations both before and immediately after a 'supermaximum' is somewhat erratic. The star field may be readily found from the 5.76 magnitude star 57 Ursae Majoris and being circumpolar in northern latitudes, the variable is observable all the year round. During the summer months, when it lies below the pole, twilight often interferes with its observation and the minima are then difficult to observe.

Joy[3] found the spectrum at maximum to be essentially continuous with an energy distribution like that of a B5 star. As the variable declines in brightness, the bright emission lines emerge and at minimum there are bright lines of the Balmer series of hydrogen of medium intensity together with very faint lines of neutral helium and a faint K line of singly ionised calcium. The absorption spectrum of the cooler component is also faintly visible around minimum. From this, Joy concluded that the blue star may be about one magnitude fainter than the red companion, giving a true visual range of about 4.5 magnitudes.

The spectrum at minimum has also been recorded by Herbig[16] using the 82-inch Crossley reflector, confirming the presence of wide emission lines of hydrogen and probably of helium also on a rather weak continuum. A spectrogram taken by the same observer when the star was near a normal maximum, visual magnitude 11.9 showed an absorption spectrum with very broad hydrogen lines which run together after H9. The K line of ionised calcium was also present, the ratio of its intensity to the Balmer lines corresponding to a spectral type of A5 or A7. The neutral helium line at λ4471 is also visible, this latter feature being more characteristic of an early B-type star than the later spectral classification suggested by the Balmer lines.

One further feature of SU Ursae Majoris which is worthy of note here is that, like SS Cygni, and especially AE Aquarii, it shows very pronounced flare-like activity when at minimum.

From photoelectric studies made in blue light, Mumford[92] has found oscillations of the order of 0.3 to 0.4 magnitude which continue even while the star is rising to a maximum. An initial analysis would suggest that these fluctuations are semi-periodic in nature with a period of about 5 minutes. There is, however, no evidence of any eclipse in this particular system.

SW URSAE MAJORIS

$08^h\ 33^m\ 09^s\ +53°\ 40'.0$ (1950.0) $10^m.6–15^m.8$ (visual)

Like SU Ursae Majoris, this variable has been known for more than sixty years and has been observed fairly comprehensively for several decades. Unlike the previous star, however, the mean period of this U Geminorum variable is certainly very long. The figure originally proposed of about 1,000 days, was certainly questionable although admittedly, the amplitude of 5.2 magnitudes is greater than that for most of these stars and from the cycle-amplitude curve we would anticipate it to have a long mean period. More recent observations indicate a somewhat shorter mean cycle of 459 days which is still the longest to have been definitely established for a U Geminorum variable, approached only by that of UV Persei.

The maxima are, in general, of short duration, and often much fainter than the extreme figure given above. We therefore have a situation similar to that found for UV Persei, namely that several maxima have, in all probability, been missed completely due to various causes, not the least of these being bad weather conditions. Bearing this in mind, it is likely that, as more observational data accumulate, the mean period of 459 days may be reduced still further. As we shall see later, however, there is some observational evidence that this particular binary system is somewhat abnormal and SW Ursae Majoris may be a link between the U Geminorum stars and the recurrent novae.

As might be expected, very large instruments are necessary to observe the star when around minimum and the light variations at this phase are virtually unknown. In the accompanying light curve (Fig. 58), the dotted portion must consequently not be construed as giving a true picture of the behaviour during this phase, merely representing the probable curve when the star is faint and beyond the reach of most instruments.

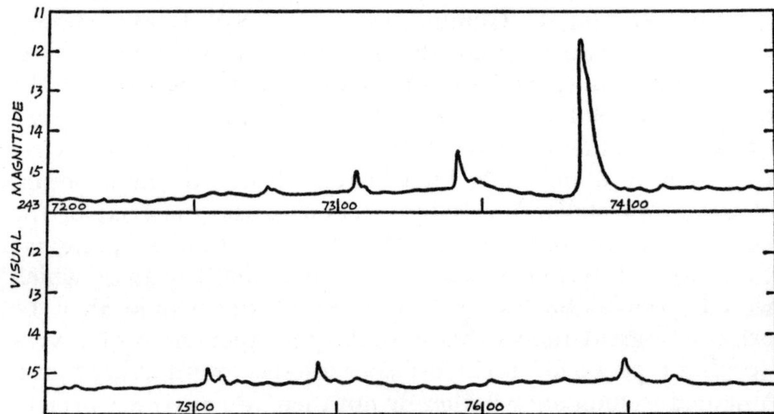

Fig. 58 – Light curve of SW Ursae Majoris (*Courtesy of the British Astronomical Association*).

One spectrogram taken by Herbig[16], again with the 82-inch Crossley reflector, when the star was near minimum with a visual magnitude of approximately 16.0, shows only weak emission lines of Hγ and Hδ on a very weak continuum. At this magnitude, the star is clearly at the very limit of spectroscopic facilities for the production of high-dispersion spectrograms.

Now it is possible to estimate the surface temperature of the white dwarf component from the spectrophotometric gradient of the star when at maximum and by this method the extremely high surface temperature of 60,000°K has been obtained. Earlier, we have seen that if we make the asumption that these stars radiate essentially as black bodies at maximum, the surface temperatures of U Geminorum and SS Cygni are 15,000°K and 12,000°K respectively so clearly the figure obtained for SW Ursae Majoris is clearly abnormal, particularly for a fairly typical white dwarf star. If the figure is correct (and at present it is difficult to see where any error has been made), then obviously the blue component of this U Geminorum system is more akin to the hot components of the recurrent nova than those of the normal dwarf novae. This observation gains some support, of course, from the extremely long mean cycle of this star and the large amplitude. On the other hand, if we were to apply the result found by Krzeminski[31] for U Geminorum to the SW Ursae Majoris system and assume that the secondary component is responsible for the typical outbursts, we immediately run into the difficulty of explaining how the surface temperature of a red dwarf can suddenly increase to 60,000°K. If we take the view that such an abnormally high increase is due to the exposure of the lower stellar levels by ejection of the material of the atmospheric regions through the inner Lagrangian point we run into very serious trouble since the amount of mass which would have to be lost in a very short time would then be extremely great unless this secondary component is of a very peculiar type. So far, no high-dispersion spectrograms have been obtained to indicate whether or not there are any lines present due to ionised helium such as are found in the recurrent novae.

BB VELORUM

$08^h\ 36^m\ 02^s$ $-47°\ 12'.3$ (1950.0) $13^m.2$–$15^m.2$ (photographic)

Discovered by de Kort [93], this variable lies in a very crowded field of faint stars making positive identification hazardous even when at its brightest. No figure for the mean cycle has been given and nothing is yet known of its spectroscopic characteristics.

CU VELORUM

$08^h\ 56^m\ 37^s$ $-41°\ 36'.3$ (1950.0) $12^m.0$–($15^m.5$ (photographic)

This variable was discovered by Hoffmeister[21] but in spite of its comparative brightness at maximum, has been relatively little observed. It lies within the southern Milky Way, close to the 4.42 magnitude star W Velorum. Here it must be mentioned that owing to some confusion at the time of the charting of this particular constellation, there are two stars which have been designated W Velorum. The star mentioned above is constant in brightness. The true variable star, W

Velorum, which is a typical long period variable with a period of 393.45 days and varying between 9.5 and 15.0 magnitude, lies some distance from this particular star field.

There are several faint stars in the neighbourhood of CU Velorum, including one very close companion, from which it must be clearly separated when at minimum for accurate visual estimates of brightness to be made. Since this U Geminorum variable has been included in the observing programme of the Variable Star Section of the Royal Astronomical Society of New Zealand, it is very probable that a reasonably complete light curve will soon become available for discussion.

TW VIRGINIS

$11^{h}\ 42^{m}\ 48^{s}$ $-04°\ 09'.3$ (1950.0) $11^{m}.8$–$16^{m}.2$ (photographic)

This U Geminorum variable has been kept under fairly constant observation (apart from those months when the sun

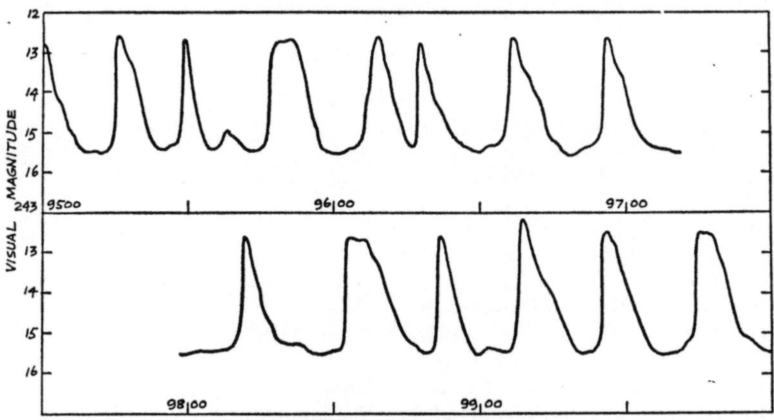

Fig. 59 – Light curve of TW Virginis (*Courtesy of the American Association of Variable Star Observers*).

passes through this zodiacal constellation) by the American Association of Variable Star Observers[7]. The mean period is quite short, being only 25 days, and the typical light variations are well illustrated in the accompanying light curve (Fig. 59).

TW Virginis lies on the border with Leo, the field being readily located from its proximity to the fourth magnitude red star ν Leonis. Both long and short maxima have been recorded, the former being typically flat. In general, the rise is characteristically steep with the decline being more protracted and often with a pronounced hump which occurs usually between thirteenth and fourteenth magnitude. The time spent at minimum varies quite considerably and fluctuations of the order of half a magnitude have been noted on occasion.

SW VULPECULAE

$19^h\ 57^m\ 56^s$ $+22°\ 47'.9$ (1950.0) $15^m.6$–($18^m.2$ (photographic)

Here we have an extremely faint U Geminorum variable which was found by Rosino[94] in 1948. Although a very difficult star to observe, there would appear to be little doubt of its type (if we exclude the possibility that it may belong to the Z Camelopardalis subgroup) and the very short mean period of 13 days has been derived from a study of the photographic plates on which it was discovered.

REFERENCES

1 Markarian, B. E., *Astrofizika*, **9**, No. 4, 511 (1967)
2 Zinner, E., *A.N.*, **265**, 345 (1938)
3 Joy, A. H., *Publ. A.S.P.*, **52**, 324 (1940)
4 Crawford, J. A., and Kraft, R. P., *Ap. J.*, **123**, 44 (1956)
5 Walker, M. F., *Sky & Tel.*, **29**, 23 (1965)
6 Shapley, H., and Hughes-Boyce, E. M., *H.A.*, **90**, 169 (1934)
7 Quarterly Reports, A.A.V.S.O.
8 Elvey, C. T., and Babcock, H. W., *Ap. J.*, **97**, 412 (1943)
9 Himpel, K., *A.N. Circular*, **26**, 25 (1944)
10 Morgenroth, O., *Publ. Berlin-Babelsburg Univ. Obs.*, **19** (1938)
11 Hoffleit, D., *H.B.*, **887** (1932)
12 Rohlfs, E., *Publ. Obs. Ger. Acad., Berlin*, **1**, No. 3 (1949)
13 Shapley, H., and Swope, H. H., *H.A.*, **90**, No. 5 (1934)
14 Swope, H. H., *ibid*, **90**, No. 7 (1936)
15 Hartwig, E., *A.N.*, **4238**, 223 (1908)
16 Herbig, G. H., *Ap. J.*, **131**, 632 (1960)
17 Kraft, R. P., *ibid*, **135**, 408 (1962)
18 Hanley, C. M., and Swope, H. H., *H.B.*, **913**, (1940)
19 Hertzsprung, E., *Bull. Astron. Inst. Netherlands*, **4**, 172 (1928)
20 Hoffmeister, C., *Publ. Berlin-Babelsburg Univ. Obs.*, **28**, (1943)
21 Hoffmeister, C., *Astron. Thesis, A.N.*, **12**, No. 1 (1949)

22 Hoffleit, D., *H.B.*, **874** (1930)
23 Bateson, F. M., *V.S.S.*, *R.A.S.N.Z.*
24 Rybka, E., *Lwow Contrib.*, **2** (1934)
25 Opalski, A., *ibid*, **4** (1936)
26 Erro, L. E., *H.B.*, **913** (1940)
27 Huruhata, M., *ibid*, **913** (1940)
28 Wells, L. D., *H.C.O. Circular*, No. 12 (1896)
29 Lortet-Zuckermann, M. C., *Ann. Astrophysics*, **29**, No. 30, 205 (1966)
30 Strand, K. A., *Ap. J.*, **107**, 106 (1948)
31 Krzeminski, W., *ibid*, **142**, 1051 (1965)
32 Mannino, G., and Rosino, L., *Padua Publ.*, No. 14 (1950)
33 Miczaika, G., and Becker, W., *Heidelberg Veröff.*, **15**, No. 8 (1948)
34 Baade, W., *A.N.*, **249**, 271 (1933)
35 Hoffmeister, C., *Publ. Berlin-Babelsburg Univ. Obs.*, **1**, No. 3 (1949)
36 Rohlfs, E., *Comm. Obs. Berlin-Babelsburg & Sonneburg*, **120**, (1950)
37 Shapley, H., and Hughes-Boyce, E. M., *H.A.*, **90**, 170 (1934)
38 Hoppe, J., *A.N.*, **254**, 369 (1935)
39 Hind, J. R., *M.N.R.A.S.*, **XVI**, 56 (1856)
40 Hind, J. R., *Publ. Radcliffe Obs.*, **XV**, 285 (1857)
41 Pogson, N. R., *M.N.R.A.S.*, **LXVII**, 119 (1883)
42 Van der Bilt, J., *Reserches Astronomiques Utrecht*, **III** (1908)
43 Copeland, L., *The Observatory*, **5**, 110 (1882)
44 Knott, G., *ibid*, **5**, 100 (1882)
45 Fleming, W. P., *H.A.*, **56**, 210 (1912)
46 Hoffmeister, C. (MS)
47 Ahnert, P., *Comm. Obs. Berlin-Babelsburg*, **24** (1941)
48 Hoffmeister, C., *Comm. Obs. Berlin-Babelsburg & Sonneburg*, **115** (1950)
49 Hughes-Boyce, E. M., *H.B.*, **917** (1943)
50 Hughes-Boyce, E. M., *ibid*, **903** (1936)
51 Hoffmeister, C., *A.N.*, **259**, 39 (1936)
52 Himpel, K., *Circular A.N.*, **25**, 101 (1943)
53 Peters, C. H. F., *A.N.*, **65**, 55 (1865)
54 Kukarkin, B. V., and Parenago, P. P., *General Catalogue of Variable Stars, Moscow* (1948)
55 Brun, A., and Petit, M., *Bull. Assoc. française Obs. étoiles var.*, **12**, 1 (1952)
56 Brun, A., and Petit, M., *Peremennye Zvezdy*, **12**, 18 (1959)
57 Metcalf, J., *A.N.*, **4191**, 260 (1907)
58 Parenago, P. P., *Peremennye Zvezdy*, **4**, 279 (1934)
59 Swope, H. H., and Caldwell, I. W., *H.B.*, **879** (1930)
60 Chernova, T. C. (MS)
61 Ahnert, P., *Comm. Obs. Berlin-Babelsburg & Sonneburg*, **66** (1944)
62 Anhert, P., *ibid*, **105** (1945)
63 Kruytbosch, W. E., *Bull. Astron. Inst. Netherlands*, **7**, 253 (1935)
64 Hertzsprung, E., *ibid*, **9**, 275 (1942)
65 Himpel, K., *Circular A.N.*, **26**, 25 (1944)
66 Hughes-Boyce, E. M., *H.A.*, **109**, No. 2 (1942)
67 Hughes-Boyce, E. M., and Huruhata, M., *ibid*, No. 4 (1942)

68 Rosino, L., *Bologna Publ.*, **4**, No. 2 (1941)
69 Hoffmeister, C., *A.N.*, **236**, 235 (1929)
70 Tsesevitch, B. P., (MS) (1947)
71 Soloviev, A. B., *Astron. J., Russian Acad.*, **79** (1948)
72 Ashbrooke, J., *A.J.*, **56**, 88 (1951)
73 Boyd, C. D., *H.A.*, **90**, 243 (1939)
74 Parenago, P. P., *Peremennye Zvezdy*, **7**, 155 (1949)
75 Dirks, W. H., *Bull. Astron. Obs. Netherlands*, **9**, 197 (1941)
76 Van Hoof, A., *Ciel et Terre*, **57**, 321 (1941)
77 Van Maanan, A., *Publ. Ger. Acad., Berlin*, **1**, No. 3 (1949)
78 Greenstein, J. L., *Ap. J.*, **126**, 23 (1957)
79 Van Maanan, A., *Publ. A.S.P.*, **38**, 325 (1928)
80 Kraft, R. P., *Science*, **134**, 1433 (1961)
81 Krzeminski, W., *Publ. A.S.P.*, **74**, 66 (1962)
82 Burbidge, G. R., *Ap. J.*, **137**, 995 (1963)
83 Himpel, K., *Circular A.N.*, **25**, 103 (1943)
84 Uitterdijk, J., *Ann. Sterrewacht te Leiden*, **20**, No. 2 (1949)
85 Swope, H. H., *H.A.*, **90**, 235 (1939)
86 Swope, H. H., *ibid*, **90**, 236 (1939)
87 Swope, H. H., *ibid*, **109**, No. 9 (1943)
88 Herbig, G. H., *Publ. A.S.P.*, **56**, 230 (1944)
89 Boyce, E. H., *H.A.*, **90**, 244 (1939)
90 Shapley, H., and Hughes-Boyce, E. M., *ibid*, **90**, 171 (1934)
91 Ceraski, M., *A.N.*, **4545** (1911); *H.C.O. Circular*, **176** (1913)
92 Mumford, G. S., *A.A.V.S.O. Abstracts*, p. 12 (April 1963)
93 De Kort, J., *Bull. Astron. Inst. Netherlands*, **8**, 173 (1937)
94 Rosino, L., *Bologna Publ.*, **5**, No. 8 (1948)

13. Z Camelopardalis Variables

The discussion of the individual dwarf novae in the previous chapter was entirely concerned with the members of the U Geminorum group. We must now extend our discussion to include the small subgroup of these variables known as the Z Camelopardalis stars. In Chapter Two where we examined the light variations of the dwarf novae, we saw that the main criterion for placing a dwarf nova in this class is the presence in the light curve of periods of 'standstill', during which the brightness of the variable remains constant to within about half a magnitude. These 'standstills' may be of any duration from only a few days to as long as eighteen months.

At the present time, very few of these variables are known and of these, some have been only provisionally assigned to this subgroup. There still remains some doubt as to their actual membership. A study of the light changes of those Z Camelopardalis variables which have been extensively observed indicates that the frequency of the 'standstills' varies quite appreciably from star to star. In some, for example Z Camelopardalis, RX Andromedae and TZ Persei, they are of fairly frequent occurrence while in others, such as AH Herculis and CN Orionis, they appear to be much less common. This immediately suggests that some of the variables at present included among the U Geminorum stars may eventually have to be transferred to this subgroup as more extended series of observations become available, this being especially so for the fainter members of the class.

As before, we are faced with the difficulty that most of these stars are extremely faint objects, even when at maximum, and for much of the time they behave in a similar manner to the U Geminorum variables. Only when a 'standstill' has been definitely recorded is it possible to be certain of the type of dwarf nova with which we are dealing. As we shall see too, their spectra are very similar so there is no spectroscopic method of differentiating between the two groups. There is some spectroscopic evidence that the radial velocity curves of the Z Camelopardalis variables are asymmetrical but at present it is possible to measure the radial velocities of only one or two of these stars and far more data must be gathered before this can be accepted as a reliable means of identification. If it can be shown beyond any doubt that all of the Z Camelopardalis variables show this kind of asymmetry, undoubtedly this will open up new avenues of research into the physical nature of the 'standstills'. At the moment, our knowledge of the basic cause of these peculiar periods of comparative inactivity is very meagre indeed.

Although one explanation of an asymmetrical radial velocity curve is that the orbit of such a binary is eccentric, unlike those of the related U Geminorum stars which appear to be essentially circular, there is an alternative explanation based upon irregularities in the motion of the gas stream in such a binary system and it is a difficult problem to decide which of these two possibilities is the correct one.

RX ANDROMEDAE

$01^h\,01^m\,45^s$ $+41°\,01'.9$ (1950.0) $10^m.3$–$13^m.6$ (visual)

Since its discovery in 1905[1], this star has been kept under constant observation and, especially in recent years, its light curve has been fairly well defined with only a few gaps. It is a typical Z Camelopardalis variable as may be seen from Fig. 60 and exhibits not only the usual 'standstills', but also periods of complete irregularity during which it is virtually impossible to define either maxima or minima with any degree of precision.

If we ignore the periods of 'standstill', the mean cycle of this star is 14.1 days which is in agreement with the much shorter periods generally found for these stars. When the star is undergoing neither a 'standstill' nor a period of irregular behaviour, the maxima are very like those of the U Geminorum variables with rapid rises to maximum and a somewhat slower decline.

The field of this variable is readily found from the sixth magnitude star 39 Andromedae and there is a very close, faint companion from which it must be clearly separated when at minimum. The star is not strictly circumpolar in British latitudes, lying extremely close to the northern horizon during the summer months. At such times the maxima may be fairly readily observed but the minima are accessible only with large instruments.

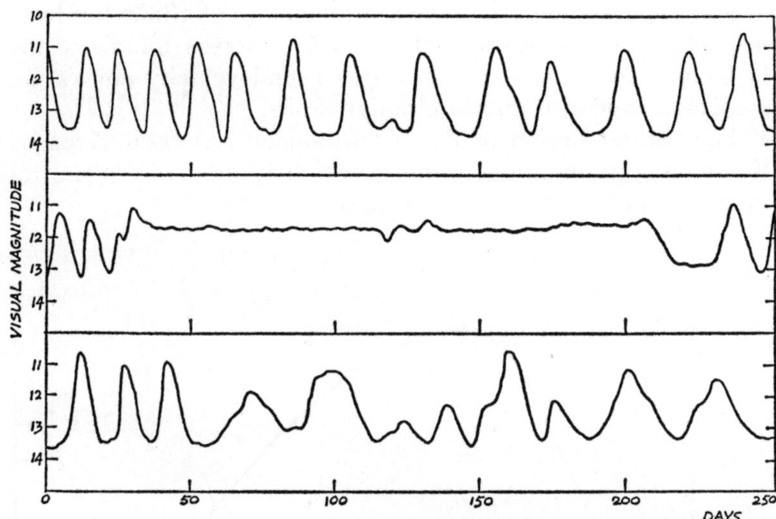

Fig. 60 – Light curve of RX Andromedae (*Courtesy of the British Astronomical Association*).

The spectrum of this star at minimum has been shown by Kraft[2] to consist of strong emission lines of hydrogen, with the Balmer jump in emission, together with somewhat weaker ones of singly ionised calcium. The lines due to neutral helium at λ4771 and λ4026 are also present in emission while the bright line of a singly ionised helium at λ4686 is just visible at this phase. From this it would seem that the primary component is, like those of the majority of the dwarf novae so far investigated spectroscopically, an sdBe type white dwarf. There is, however, no evidence of an absorption spectrum and so we know nothing of the spectral class of the secondary.

The radial velocity of the white dwarf primary as measured from the emission lines varies between +130 and −225 kilometres per second and it is probably significant that the radial velocity curve (Fig. 61) is highly asymmetrical indicating a high eccentricity of the orbit. Although we must not overlook the possibility that the radial velocity curve is actually symmetrical yet in some way distorted from a sine wave due to the motion of the gas stream in this system, the probability is that this asymmetry of the curve does represent an eccentricity of the orbit, particularly since a similar stare of affairs has been suspected for Z Camelopardalis itself[2] whereas for all of the U Geminorum variables for which radial velocity curves are available appear to possess circular orbits.

The orbital period of RX Andromedae has been given by Kraft[2] as 5 hours and 5 minutes which is only a little shorter than that of Z Camelopardalis itself.

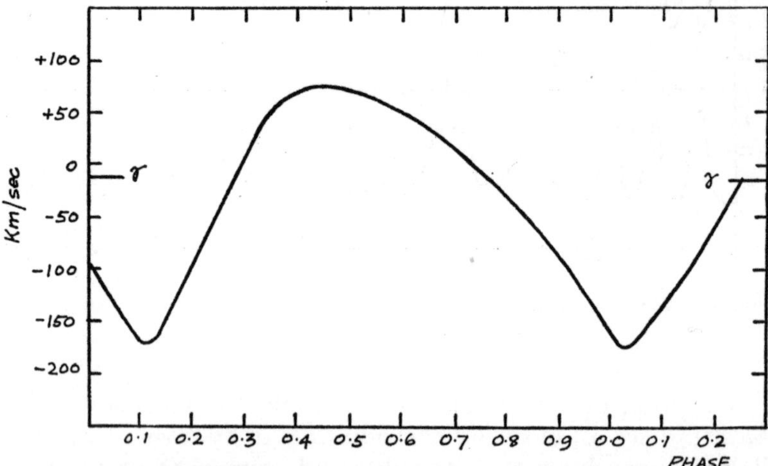

Fig. 61 – Radial velocity curve of RX Andromedae (*Astrophysical Journal by kind permission of the University of Chicago Press*).

As we have seen earlier, those dwarf novae which have short orbital periods show no sign of the absorption spectrum of the secondary which in this case would appear to be relatively fainter than the blue primary. Relatively few studies of the spectrum of this star during a 'standstill' have been made. Elvey and Babcock[3] have, however, investigated the spectral changes which occur at this phase while the brightness was varying between 11.5 and 12.3 magnitude. Throughout the whole of this period, bright lines of hydrogen are visible but no lines due to helium were recorded. Certain other important features were also noted by these observers. For example, during those times when the brightness at 'standstill' was relatively faint, namely about 12.3 magnitude, a faint emission line in the position of the K line of singly ionised calcium was recorded and the other lines in the spectrum were at their most intense. They were also more numerous at this phase.

In addition, the colour of the star varies. During 'standstill' it has been noted as very similar to that of an A7 or A8 star, somewhat later than that expected from a typical white dwarf. As we saw during the discussion of the nature of the 'standstills' of this class of variable, it is quite possible that fairly marked changes occur in the cooler companion and it may be that these are responsible for the colour difference at 'standstill' rather than any change in the sdBe dwarf companion.

V800 AQUILAE

$18^h\ 54^m\ 10^s$ $+10°\ 44'.3$ (1950.0) $14^m.0$–$16^m.0$ (photographic)

The variable, V800 Aquilae, was discovered as recently as 1949 by Hoffmeister[4]. Like most of the Z Camelopardalis stars it has quite a small amplitude although the photographic range of only two magnitudes is at the lower end of the amplitude scale even for this class of star. It must be borne in mind, however, that it has been observed only intermittently since its discovery and as further observational data accumulates, the above range may have to be revised.

The star lies just within the Milky Way, in the same field as as NGC 6724. Unfortunately, it also lies within a very densely populated field of faint stars and as a result it is not an easy object to identify even when at maximum and one must guard against mis-identification. At present, it has not been sufficiently observed for a mean period to be assigned. Four maxima have been recorded by the author during 1967–68 together with a 'standstill' lasting for approximately two months (Fig. 62), with the star remaining at about 14.7 magnitude with only minor fluctuations.

From the light curve given in Fig. 62, it will be seen that the rises are quite rapid, resembling Z Camelopardalis and like this star, the onset of a 'standstill' would appear to be, from the limited observational data available, about a third of the way down from maximum to minimum brightness. No spectroscopic details are yet available for this particular variable. When at minimum, it will be close to the limit of detection.

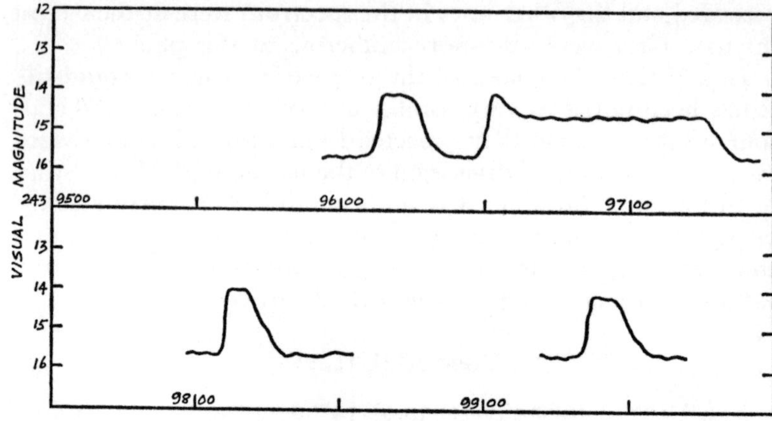

Fig. 62 – Light curve of V800 Aquilae.

Z CAMELOPARDALIS

$08^h\ 19^m\ 43^s$ $+73°\ 16'.6$ (1950.0) $10^m.2$–$13^m.4$ (visual)

Z Camelopardalis is another variable which has been quite fully observed, both visually and spectroscopically, since its discovery in 1904[5] and was the first of this small subgroup to be differentiated from the main class of dwarf novae. Like the other members of this group which have been comprehensively observed, we may define three characteristic features of the light curve (Fig. 63); periods of normal behaviour when the star exhibits light variations which are virtually indistinguishable from those of the more abundant U Geminorum variables, the 'standstills' when the light remains almost constant, and times of completely erratic light fluctuation when the star varies irregularly, seldom attaining either normal maximum or minimum brightness.

At those times when the star is behaving more or less normally, the maxima occur quite close to the mean period of about 23 days with the decline being generally much slower than the rise. Any deviations from the mean cycle are usually quite small. At such times too, the minima are reasonably undisturbed, almost invariably around 13.3 magnitude and only minor fluctuations of the order of 0.2 magnitude have been recorded. From time to time, of course, this sequence of

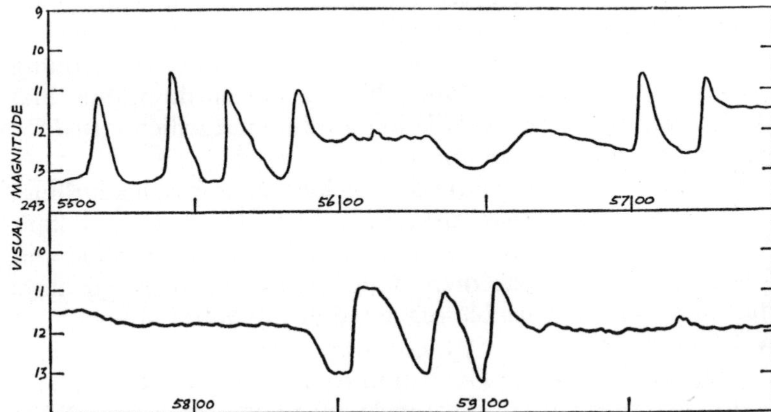

Fig. 63 – Light curve of Z Camelopardalis (*Courtesy of the British Astronomical Association*).

events is interrupted by the characteristic 'standstills' which, as far as Z Camelopardalis is concerned, occur during the fall to minimum at about magnitude 11.2 with variations of about 0.5 magnitude taking place at irregular intervals. As with all of the Z Camelopardalis stars which have been thoroughly observed, no trace of any periodicity has been found for the minor variation in brightness during a 'standstill'.

Coming now to the third feature of this star, long periods of irregular behaviour are much less frequent than the 'standstills' but recently, from 1956 to 1959, this variable underwent a truly remarkable series of fluctuations, almost unique in the history of this star since comprehensive observations commenced at the beginning of the century[6]. Fortunately, the variable was extensively observed throughout the whole of this period, particularly by the Variable Star Section of the British Astronomical Association[6].

Since these observations give us an excellent insight into the behaviour which can be expected of these variables, it is worthwhile describing them in some detail here. Before the summer of 1956, there was little indication of the highly abnormal behaviour which was to come apart from a 'standstill' beginning on April 27 of that year and lasting for 31 days. Normal light variations were then resumed but in the September, however, a second 'standstill' began which continued for 105 days. Although March, 1957 saw a normal maximum, the star rose almost immediately from the ensuing minimum to a further 'standstill' of about 80 days, one final curious feature of that year being a minimum which lasted for 32 days, far longer than normal.

During the following year Z Camelopardalis was, if anything, even more erratic than previously. An abnormal and faint maximum on February 12 attained only 10.9 magnitude and followed another extremely long period of over 50 days during which the variable remained at minimum although, as will be seen from Fig. 63 there were considerable fluctuations during this minimum. The extremely irregular light variations continued throughout the whole of 1958 with most of the maxima being well below average brightness and interspersed by highly disturbed minima. In 1959, however, the 'standstill' which had commenced in the previous December ended, not in a fall to minimum as had always happened before, but in a rise to maximum! Such an occurrence remained unique among the Z Camelopardalis variables until September 1968 when a similar rise from a 'standstill' was recorded for TZ Persei.

From the foregoing account, it will be readily appreciated that continued accurate visual observation of this, and all other dwarf novae, is vital for our understanding of these peculiar variables. Like the 'standstills', the onset of such periods of intense activity is completely unpredictable.

The star is circumpolar and readily accessible all the year round in northern latitudes although twilight often interferes during the summer months when it lies fairly close to the horizon below the pole. The field is fairly open and is readily found from F1 22 and F1 27 Ursae Majoris. There are also four faint comparison stars in the neighbourhood of the variable which are useful for estimating its magnitude when at minimum.

The spectrum of Z Camelopardalis at minimum has been observed by Elvey and Babcock[3]. The continuum is then similar to that of a dG5 star although no absorption lines were found during their investigation, their spectrograms being taken with quite a low dispersion. Several fairly strong Balmer lines are present in emission together with a few weak neutral helium lines and a slightly more intense K line of singly ionised calcium. All of these bright lines are wide and somewhat diffuse suggestive of a high rotational velocity. As the star brightens to a maximum, the emission lines weaken and are replaced by broad, shallow absorption lines of hydrogen. With declining light after a typical maximum, the spectrum appears continuous, probably because of the gradual emergence of the emission lines which, for a time, obliterate the absorption spectrum and finally reappear fully as minimum is approached.

Relatively few spectrograms of this variable during a 'standstill' have been obtained. Those which are available show that it then closely resembles a typical U Geminorum star during the decline from maximum. One remarkable feature is the form of the absorption lines during this phase, these being broad and approximately symmetrical with greatly extended wings about a central emission.

SY CANCRI

$08^{h}\ 58^{m}\ 13^{s}$ $+18°\ 06'.2$ (1950.0) $10^{m}.9$–$13^{m}.8$ (photographic)

The irregular light variations of this star were first noted by Jacchia[7] who assigned the variable to the class of RW Aurigae stars. These are curious variables which show rapid and completely irregular changes in their light curves and are normally found in great numbers in what are known as the T-associations in the constellations of Auriga, Taurus, Orion and Monoceros. Their spectra indicate that they are mostly dwarf stars of spectral type G and bright emission lines are a common feature of their spectra. It is not difficult to see, therefore, that on the basis of the spectrum alone, it would be easy to confuse SY Cancri with one of these variables.

More detailed investigations of this star have been carried out recently, particularly by Herbig[8], which suggest that it should be placed in the Z Camelopardalis class. The light curve (Fig. 64), drawn from 406 observations made by the author during 1967–68, would tend to support this latter classification.

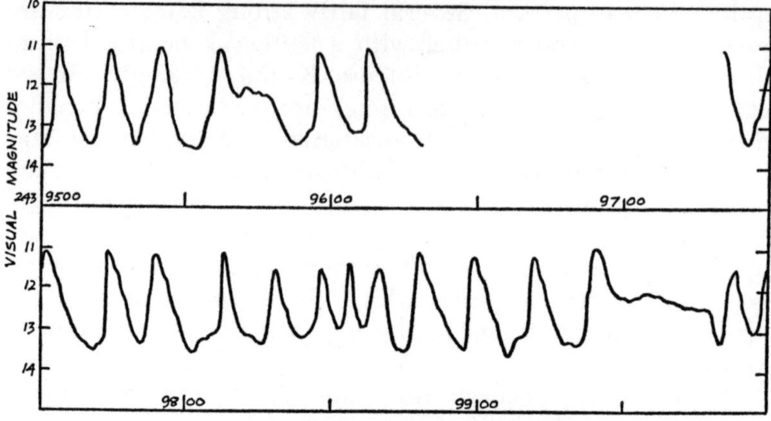

Fig. 64 – Light curve of SY Cancri.

As will be seen, even during this limited period, the star exhibited quite erratic behaviour on several occasions and from this, it would seem that SY Cancri was undergoing one of these periods of irregular light fluctuations when observed by Jacchia.

SV CANIS MINORIS

$07^h\ 28^m\ 27^s$ $+06°\ 05'.1$ (1950.0) $13^m.0$–$16^m.3$ (photographic)

SV Canis Minoris is yet another of the many variable stars of all classes discovered by Hoffmeister[9]. The field is readily found from the 5.34 magnitude star η Canis Minoris and is not far from the loose star cluster H VIII 44 which was discovered by Sir William Herschel. As with many of these variables, there is a faint companion star at position angle 310° from which the variable must be clearly separated, especially when at minimum, quite large apertures being necessary for the observation of this star during this phase. There is still some dubiety about any 'standstills' of SV Canis Minoris and it is not yet certain that it is a true Z Camelopardalis variable.

Owing to a lack of observations made over a long period, the light curve is still very poorly defined and no mean period can be assigned to it. The spectrum too, which would provide us with useful information has not been studied in sufficient detail to remove this uncertainty concerning its type. It is quite possible that the star may be a faint U Geminorum variable. On the other hand, it may even be a semi-regular variable with a reasonably short period very similar to VZ Camelopardalis. Being a winter star in northern latitudes, there are long periods during the summer when it is unobservable.

BS CEPHEI

$22^h\ 27^m\ 27^s$ $+64°\ 59'.3$ (1950.0) $14^m.0$–$16^m.0$ (photographic)

This particular Z Camelopardalis variable was discovered in 1941 by Hoffmeister[10] during an extensive photographic survey of this constellation which resulted in the discovery of thirty variable stars of all types. The photographic amplitude, like that of V800 Aquilae, is quite small, only two magnitudes,

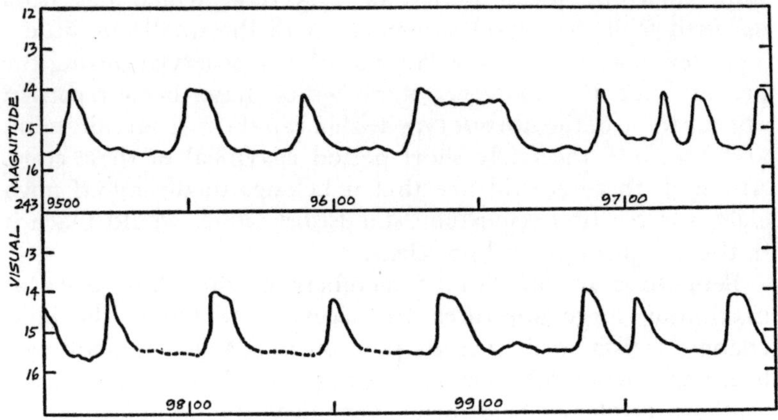

Fig. 65 – Light curve of BS Cephei.

while the mean period given in the literature of approximately 40 days is more like that of a U Geminorum variable than one of the Z Camelopardalis subgroup.

The light curve (Fig. 65), drawn from 377 observations made during 1967–68 using a 13-inch reflector, yields a mean period of about 34 days.

The star lies not far from the 5.66 magnitude star 26 Cephei and is again a circumpolar variable in northern European latitudes. The visual range would appear to be similar to the photographic, varying from 13.8 to 15.7 magnitude. Only one short 'standstill' was recorded during the time the star was under observation, the brightness at this phase being 14.4 magnitude with only minor fluctuations of the order of 0.2 magnitude. As will be seen from the light curve, there is a tendency for bright and faint maxima to alternate. Quite clearly, BS Cephei is a star for which many more observations, using large instruments, are required before these anomalies can be fully resolved.

BP CORONAE AUSTRINUS

$18^h\ 33^m\ 26^s$ $-37°\ 28'.4$ (1950.0) $13^m.9$–$15^m.6$ (photographic)

One of the probable members of the Z Camelopardalis group, this variable was found as long ago as 1931 by Gerasimovic[11]. If this star obeys the cycle-amplitude relationship, then the mean period of 13.5 days which has been assigned to it is in good agreement with the small amplitude. Very few maxima have been satisfactorily observed during the period since its discovery; those which have been recorded appear to be of the abrupt type with a rapid rise from minimum and certainly the fairly short period is typical of these stars. Although there is evidence that it belongs to the dwarf nova class, the position regarding 'standstills' which would place it in this subgroup, is still not clear.

Being one of the fainter members of the class, even at maximum, large apertures are necessary to follow the light changes satisfactorily and its position makes it a difficult star to observe from the northern hemisphere. It lies close to the southern border of Sagittarius, very near the boundary of the Milky Way in a rich field of faint stars.

HN CYGNI

$19^{h}\ 31^{m}\ 41^{s}$ $+28^{\circ}\ 49'.3$ (1950.0) $13^{m}.8$–$15^{m}.9$ (photographic)

This peculiar variable has been tentatively assigned to the Z Camelopardalis group but the number of observations which have been made are totally inadequate to define the light curve sufficiently to resolve the question of its true nature, the light variations as shown in Fig. 66 appear to be completely irregular, rather reminiscent of an RW Aurigae variable.

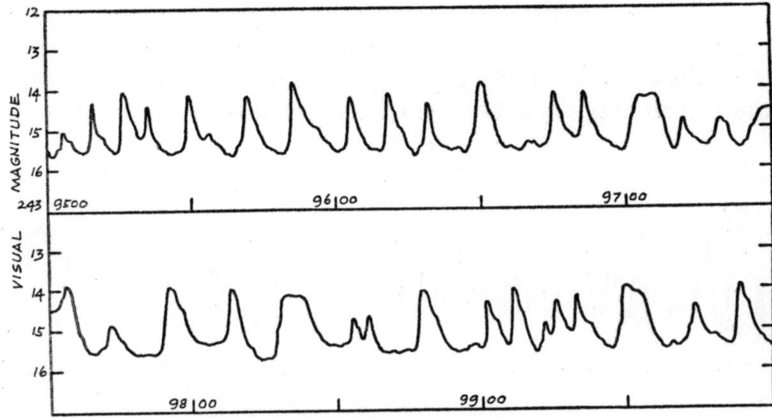

Fig. 66 – Light curve of HN Cygni.

It has to be pointed out, however, that this star was under observation for only a comparatively short period and it is faint at all phases of its light variation. It may also be that during this period it was undergoing a period of erratic behaviour which is a characteristic feature of these variables.

AB DRACONIS

$19^{h}\ 51^{m}\ 07^{s}$ $+77^{\circ}\ 37'.1$ (1950.0) $12^{m}.0$–$15^{m}.5$ (visual)

AB Draconis has been known for almost half a century but being somewhat fainter than either RX Andromedae or Z Camelopardalis, it has only been extensively observed during the past two decades or so, mainly by the American Association of Variable Star Observers[12]. The 'standstills' of this star appear to be of less frequent occurrence than in either of the two former stars and certain other features are apparent from the light curve.

Apart from the 'standstills', there are two types of maxima. In the first, the star brightens to a very sharp peak, remaining at maximum for only one or two days before declining somewhat more slowly. The second, rarer, type has a broad, flat maximum and in general, these maxima are interspersed by five or six of the normal type. Quite often, too, there are humps on the descending branch of a maximum, any such irregularities on the ascending branch being confined almost exclusively to the broad type maxima. It will also be noted that the minima vary quite appreciably and over certain periods there even appears to be a wave associated with the minima.

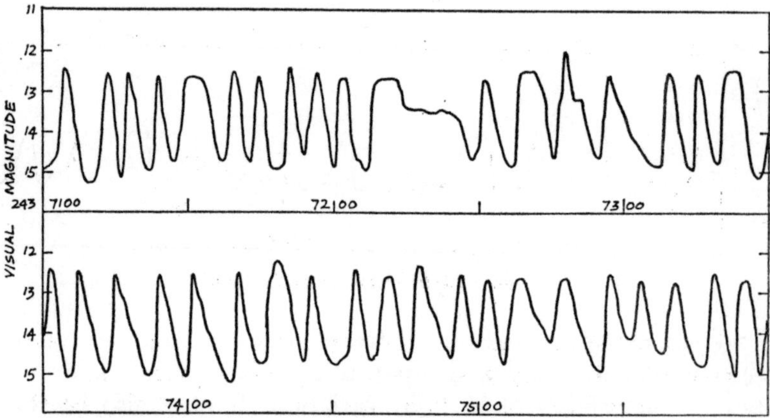

Fig. 67 – Light curve of AB Draconis (*Courtesy of the American Association of Variable Star Observers*).

AB Draconis is one of the most northerly of the dwarf novae, being readily found from the fourth magnitude star κ Cephei. Being circumpolar it is readily observable all the year round from the northern hemisphere and even at its most inaccessible, it is relatively high above the horizon. As a result, very few maxima have been missed in recent years but the minima can be observed only with moderately large instruments, especially during the summer months when twilight interferes with observation. There are several very faint stars in the field of the variable making positive identification extremely difficult when around minimum. This is particularly the case for visual observers, especially when averted vision has to be employed.

The spectrum of this star when at maximum has been examined under low dispersion by Elvey and Babcock[3]. It then consists of broad, shallow absorption lines of the Balmer series. At minimum phase when the variable was at 15.3 magnitude, Herbig[13] recorded very strong, narrow emission lines of hydrogen and singly ionised calcium on a hot continuum. The emission lines of neutral helium, although present, were extremely faint.

AH HERCULIS

$16^h\ 42^m\ 05^s$ $+25°\ 20'.5$ (1950.0) $10^m.9$–$13^m.9$ (visual)

Although this variable has not been as attentively studied as, for example, RX Andromedae and Z Camelopardalis, the light curve as shown in Fig. 68 is fairly complete over the last few years with only one or two gaps which are most noticeable during the early months of the year when the star does not rise until the early hours of the morning. Only a few 'standstills' have been recorded during the past decade or so, none of any great duration. Certainly there have been no lengthy 'standstills' such as have been recorded for RX Andromedae, Z

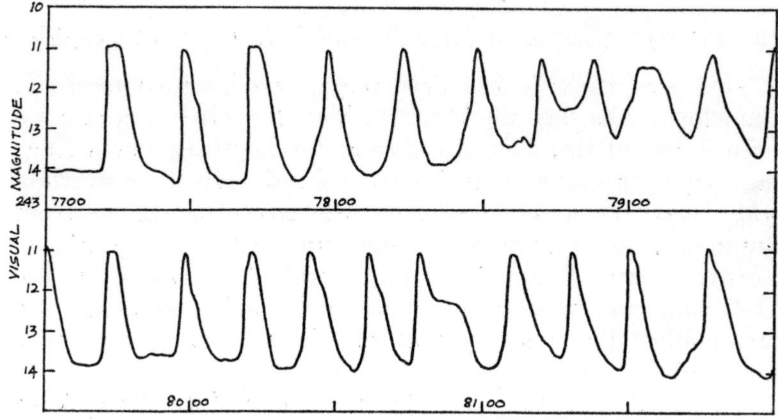

Fig. 68 – Light curve of AH Herculis.

Camelopardalis and TZ Persei over the past few years. The onset of a 'standstill' occurs about a third of the way down the descending branch of the light curve, AH Herculis being like Z Camelopardalis in this respect.

An account of the early observations of this star has been given by Jacchia[14] covering the period up to 1941. As with almost all of these variables, the photographic amplitude of 10.9 to 14.7 magnitude is greater than the visual range. The colour index of this star at minimum deserves some comment here. For these variables, taken as a whole, the colour indices at minimum range from +0.2 to +0.5 magnitude. A colour index as high as +0.8 magnitude is somewhat unusual for a dwarf nova but it should be noted that certain minima as deep as 14.2 magnitude (visual) have been recorded and we must therefore accept the minimum visual magnitude of 13.9 with some reserve.

The mean cycle length of 19.6 days is in good agreement with the cycle-amplitude relationship. Again, as with many of these variables, two types of maxima are known, long and short, and there is a tendency for the latter type to exhibit marked irregularities on the descent to minimum.

The field of AH Herculis is readily located from the fifth magnitude star 51 Herculis and is fairly open with few very close companions to complicate observation.

KW HERCULIS

$18^h\ 39^m\ 07^s$ $+12°\ 06'.4$ (1950.0) $13^m.8$–$16^m.5$ (photographic)

This faint variable has been tentatively assigned to the Z Camelopardalis class by Ahnert[15]. Very few observations have been made of this star, insufficient for anything but a fragmentary light curve to be drawn and although those maxima which have been recorded show the typical steep rise from minimum, the evidence for any 'standstill' is very flimsy. Further observational data may eventually place it in the U Geminorum class rather than in this small subgroup. So far, neither the mean period nor its spectrum have been recorded.

The star lies on the very edge of an arm of the Milky Way on the borders with Aquila and Ophiuchus. There are many faint stars in the field, making identification extremely troublesome even under the most favourable conditions.

BT LACERTAE

$22^h\ 22^m\ 21^s$ $+55°\ 56'.2$ (1950.0) $12^m.6–15^m.0$ (photographic)

BT Lacertae is a further variable about which there is still some doubt as to its actual membership of the Z Camelopardalis class. It lies in the same field as the irregular variable RW Cephei (photographic range 8.4 to 9.1 magnitude; spectral class Mo).

At present we know nothing of its mean period and no 'standstills' of any length have been recorded with certainty. Like the preceding star, it may be that future observational work, carried out over a long period, will prove the star to be either a typical U Geminorum variable of a member of one of the other short-period classes. BT Lacertae is difficult to observe when around minimum since there are several faint companion stars in the field.

BI ORIONIS

$05^h\ 21^m\ 16^s$ $+00°\ 57'.9$ (1950.0) $13^m.2–16^m.6$ (photographic)

BI Orionis was discovered many years ago in 1923 by Hoffmeister[16] but it is only during recent years that it has come under the close scrutiny of variable star observers and sufficient observations have become available for a reasonably complete light curve to be drawn. There are, unfortunately, unavoidable gaps during the summer months but since the maxima occur fairly frequently in a mean period of about 25 days, a large enough number have been recorded for the general characteristics of this Z Camelopardalis variable to be known quite fully.

The light curve (Fig. 69) illustrates the typical behaviour of this star. Long and short maxima are known and in addition, certain of these show features very similar to the anomalous maxima exhibited by SS Cygni, the rise to maximum being relatively prolonged and sometimes showing quite marked irregularities. In general, the long maxima are slightly brighter than the short variety. The one 'standstill' recorded during the period the variable was under observation occurred about halfway down the descending branch of the curve.

It will be recalled that when discussing the general light variations of the Z Camelopardalis variables, it was mentioned that the frequency of 'standstills' appears to vary from one star to another. As far as BI Orionis is concerned, there is the difficulty that for certain periods, the variable is not readily observable and strictly speaking, we ought not to make a direct comparison with such stars as Z Camelopardalis, RX Andromedae and TZ Persei for which there are few, if any, gaps in their light curves over the past two decades or so. The available evidence would appear to suggest that these periods of relative inactivity are somewhat less frequent in the case of this particular variable. It would be unwise, however, to be dogmatic on this point until a much greater number of observations have been accumulated. At present we are working with far too little information.

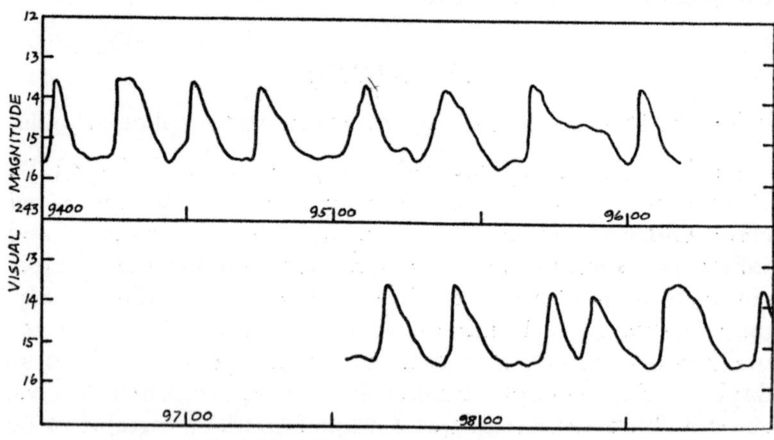

Fig. 69 – Light curve of BI Orionis.

The field of BI Orionis may be found without much difficulty from the fifth magnitude star 25 Orionis. There are two close companions of about sixteenth magnitude which prove troublesome with large instruments when the star is around minimum brightness.

Few details of the spectrum have been published. At minimum, the variable is probably at the very limit of spectroscopic determination while around maximum there are shallow absorption lines of hydrogen such as are found, for example, in Z Camelopardalis itself.

CN ORIONIS

$05^h\ 50^m\ 30^s$ $-05°\ 26'.0$ (1950.0) $11^m.8$–$14^m.7$ (visual)

Being significantly brighter than the preceding variable throughout the whole of its light cycle, CN Orionis has been assiduously observed now for many years, particularly by the American Association of Variable Star Observers[12] and the Variable Star Section of the Royal Astronomical Society of New Zealand[17]. As in the case of Z Camelopardalis, three distinct phases may be distinguished in the light curve (Fig. 70); namely normal behaviour during which the typical maxima

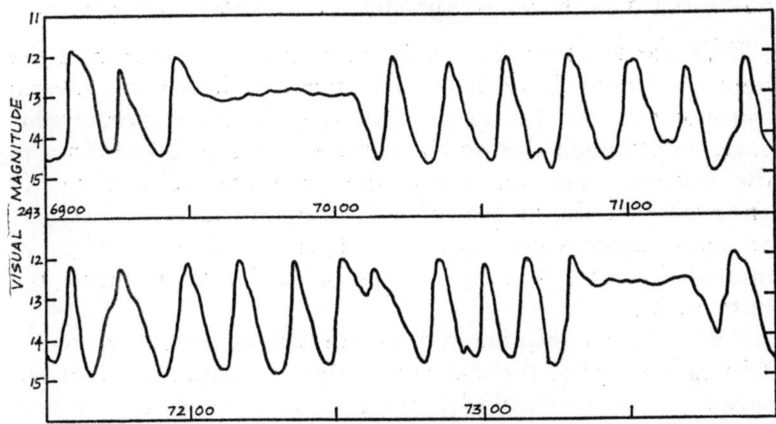

Fig. 70 – Light curve of CN Orionis (*Courtesy of the British Astronomical Association*).

occur in a mean period of 19.2 days, the 'standstills' when the star fluctuates about the thirteenth magnitude and periods of more erratic light variation. It is generally found that when the star is undergoing this irregular behaviour, the brightness rarely falls below 14.0 magnitude.

The minima of this star appear to be more variable than in the majority of Z Camelopardalis stars which have been investigated and in this respect it is interesting to compare those which occurred about JD 2437125 and JD 2437150.

The variable lies quite close to the Milky Way in a densely populated field and moderately large apertures are necessary for its visual observation when at minimum. The spectrum at this phase of the light variation is somewhat similar to that of AB Draconis, consisting of fairly narrow emission lines of hydrogen and singly ionised calcium on a hot continuum. When at maximum brightness, the broader absorption lines of hydrogen which are a general feature of these variables are visible. No spectroscopic details are yet available of this star during a 'standstill'.

TZ PERSEI

$02^h\ 10^m\ 18^s$ $+58°\ 08'.9$ (1950.0) $12^m.1$–$15^m.4$ (visual)

This star has been known for more than fifty years and the light variations have been attentively followed by several astronomers. In European latitudes, it is a circumpolar variable although during the summer months it lies close to the northern horizon below the pole and is not readly observable. Twilight too, at this time of the year, hampers accurate observation and indeed it is often invisible at minimum except in a clear, dark sky and with a fairly large instrument.

The field is easily found from the bright double cluster in Perseus but unless the star is at, or near, maximum, identification is extremely difficult owing to the presence of two faint companions, one of the fourteenth magnitude and the other an irregular variable, possibly of the RW Aurigae type, with an amplitude of about half a magnitude.

The 'standstills' of this Z Camelopardalis variable (Fig. 71) occur about a third of the way down from maximum to minimum brightness, around 13.1 magnitude, and have a tendency to be somewhat less frequent than those of Z Camelopardalis and also to be of rather long duration. At least one 'standstill' has been recorded which lasted for approximately eighteen months[18] with only minor fluctuations throughout the whole of this period and, as has been mentioned earlier, TZ Persei has also exhibited the almost unique features of ending a 'standstill' with a rise to maximum and also, in December, 1968, rising from a minimum to a 'standstill'[19]. The typical maxima of this star may be loosely classified into long and short types, with the former often being smoothly rounded at the peak. Unlike most of these variables, however, there appears to be a fairly smooth gradation between these two extreme types as will be seen from Fig. 71. Periods of more irregular behaviour tend to occur almost immediately following a long 'standstill'. During these periods, the star is rarely fainter than fourteenth magnitude.

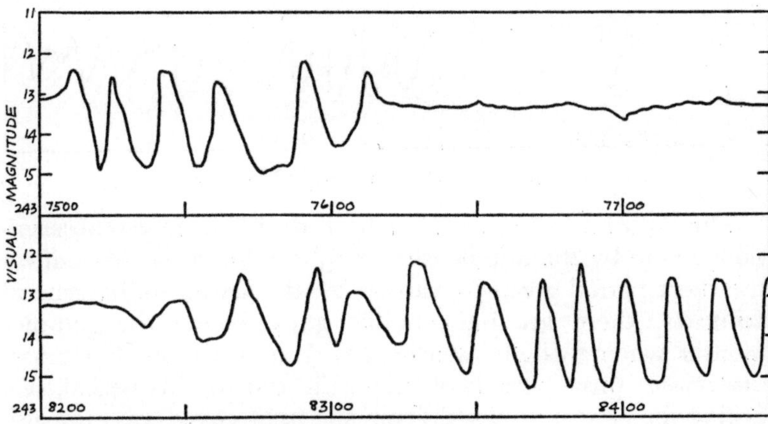

Fig. 71 – Light curve of TZ Persei (*Courtesy of the British Astronomical Association*).

Two unwidened spectrograms of TZ Persei, taken by Herbig[13] using the 82-inch Crossley reflector when the visual magnitude was 13.9 and 13.8 magnitude, show neither absorption nor emission lines, only a very hot continuum. At the present time, we know nothing of the orbital period of this variable and no photoelectric work has been carried out to determine whether or not there is any evidence of eclipse in this system.

FO PERSEI

$04^h\ 04^m\ 44^s$ $+51°\ 06'.9$ (1950.0) $13^m.8$–$16^m.2$ (photographic)

This faint Z Camelopardalis variable has been described by van de Vorde[20] who has assigned a mean period as short as 11.3 days to it. The field is readily found from the fourth magnitude star λ Persei.

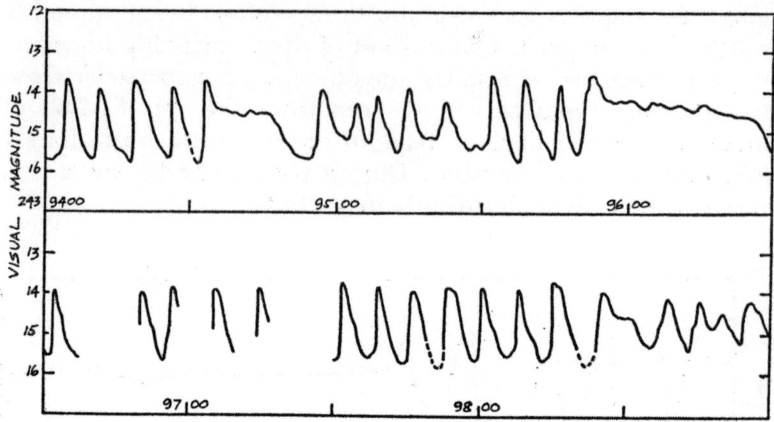

Fig. 72 – Light curve of FO Persei.

The light curve (Fig. 72) has been drawn from 390 observations made by the author during 1967–68 and would confirm the short period given by van de Vorde. Owing to its general faintness, there are unavoidable gaps during the summer months when twilight interfered with observation. FO Persei underwent three periods of 'standstill' during this period, one lasting for almost 70 days. From the light curve, we may also see that period of irregular behaviour followed two of the 'standstills', periods during which the amplitude of the maxima was reduced almost to half its normal range.

BX PUPPIS

$07^h\ 52^m\ 09^s$ $-24°\ 11'.6$ (1950.0) $13^m.8$–$15^m.8$ (photographic)

BX Puppis was discovered by Dirks[21] in 1941 during the same extensive photographic survey of this constellation which resulted in the finding of the U Geminorum variable BV Puppis mentioned earlier in the previous chapter. It lies within a very crowded field within the boundaries of the Milky Way less than half a degree from the compact cluster H VII 10. Like most of these variables, the mean period is comparatively short, being one of only 18 days, and the photographic amplitude of 2.0 magnitudes is similar to that of V800 Aquilae.

V735 SAGITTARII

$17^h\ 56^m\ 39^s$ $-29°\ 33'.8$ (1950.0) $13^m.5$–$16^m.5$ (photographic)

This particular Z Camelopardalis variable was found on plates taken by Swope[22] during a photographic examination carried out in 1940. The mean period is one of 25 days, the star being similar to Z Camelopardalis in this respect.

Once again, we are faced with the difficulty of a variable situated within a rich field of star inside the boundaries of the Milky Way and the task of observing this variable when at minimum, mentioned earlier in the case of TZ Persei, is even more acute in this instance since there are no fewer than seven faint stars all within a few seconds of arc of the variable.

REFERENCES

1 Graff, G. von., *A.N.*, **4289** (1905)
2 Kraft, R. P., *Ap. J.*, **135**, 408 (1962)
3 Elvey, C. T., and Babcock, H. W., *ibid*, **97**, 412 (1943)
4 Hoffmeister, C., *Circular A.N.*, **12**, No. 1 (1949)
5 Van Biesbroek, D., *Ann. Obs. Belg.*, (2) **13**, 21 (1904)
6 Holborn, F. M., *J. Brit. Astr. Assoc.*, **70**, No. 3, 134 (1960)
7 Jacchia, L., *Obs. Circular, A.N.*, **18**, 33 (1936)
8 Herbig, G. H., *Publ. A.S.P.*, **62**, 211 (1950)
9 Hoffmeister, C., *A.N.*, **238**, 37 (1930)
10 Hoffmeister, C., *Comm. Obs. Berlin-Babelsburg*, **24** (1941)
11 Gerasimovic, B. P., *Circular Pulkowa Obs.*, No. 9 (1933)
12 Mayall, Mrs. M. W., *Quarterly Reports, A.A.V.S.O.*
13 Herbig, G. H., *Ap. J.*, **131**, 632 (1960)
14 Jacchia, L., *H.B.*, **915** (1941)
15 Ahnert, P., *Comm. Obs. Berlin-Babelsburg*, **28** (1943)
16 Hoffmeister, C., *A.N.*, **218**, 316 (1923)
17 Bateson, F. M., *Quarterly Reports*, Variable Star Section, R.A.S.N.Z.
18 Glasby, J. S., *J. Brit. Astr. Assoc.*, **76**, No. 5, 351 (1966)
19 Glasby, J. S., *J. Brit. Astr. Assoc.*, **77**, No. 6, 436 (1967)
20 Van de Vorte, A. (MS)
21 Dirks, W. H., *Bull. Astr. Obs. Netherlands*, **9**, 197 (1941)
22 Swope, H. H., *H.A.*, **109**, No. 1 (1940)

14. Observation of the Dwarf Novae

Most of the variable stars known may be classed either as strictly periodic or semi-periodic. By this we mean either that their light variations are repeated faithfully from one maximum to the next as in the case of the eclipsing binaries of the Algol, β Lyrae and W Ursae Majoris types, or the pulsating stars such as the RR Lyrae and Cepheid variables; or there are irregularities associated with each individual cycle which show up in the shape, maximum brightness attained and the interval from one maximum to the next. Included in this latter group are, of course, the well known long period variables and the semi-regulars.

Once a complete cycle of one of the strictly periodic variables has been observed, we may be reasonably certain that in the majority of cases (there are exceptions, naturally, as to every rule, but these are not numerous), the next cycle will be virtually identical. Any small deviations, either in the period or in the form of humps on the ascending and descending branches of the light curve can be detected only by means of sensitive photoelectric techniques which are beyond the reach of most amateur observers. As a result, these variables are not well suited to a communal observing programme, particularly as extended series of observations are not necessary to define their general behaviour.

The long period variables and semi-regulars are in a somewhat different category since no two maxima are exactly alike and their amplitudes, especially those of the former type, are usually quite large, between six and twelve magnitudes, and their light changes are sufficiently slow for any scatter due to personal error to be easily smoothed out. We shall see later that there are, however, certain problems associated with their observation which add to the difficulties of making accurate visual estimates of brightness.

When we come to consider the dwarf novae, we find that they fit into neither of the above two classes. Firstly, their light changes are completely irregular. At various times throughout the book it has been mentioned that although mean periods have been given for a small number of these stars, there are very wide deviations from this mean figure (indeed the definition of this mean cycle is that it must be derived from as continuous a light curve as possible over a period of not less than a decade or so). In addition to the unpredictability of the outbursts, the light variations are often extremely rapid, a rise of five or six magnitudes in one or two days being quite commonplace. Nevertheless, in spite of the limitations which these conditions place upon visual observations, it cannot be denied that such series of estimates have provided us with almost all of the information we now possess concerning the major outbursts of these stars.

This in no way diminishes the great value of the observations which have been made comparatively recently using spectroscopic and photoelectric methods, the latter in combination with three-colour photometry, and the largest instruments available in the observatories of the world. This, of course, is a programme of research which can only be carried out using the highly refined techniques in the hands of the professional worker. It may be said that whereas the amateur observer using only modest equipment has put forward the pertinent questions by the production of fairly complete light curves over many decades, the professional astronomer is now providing the answers. Clearly, there is still an urgent need for both kinds of observation to be carried on into the future.

Although many of the dwarf novae, particularly the fainter members of the class, have been discovered by means of photography, often during surveys of the selected Milky Way fields designed for the discovery of other types of variable stars, the light curves we have for discussion are almost all based purely upon visual estimates. The reason for this is obvious. The larger telescopes in the professional observatories are all fully committed to programmes for which they alone are suited and this does not include following these variables on every possible occasion for the purpose of producing light curves. This rests almost entirely in the hands of the many thousands of amateurs. In a book of this kind, it is therefore appropriate now to examine the methods used for making such visual estimates and also to point out the various difficulties which are often encountered. Naturally, if the results obtained are to be of any value, they must be as accurate as possible and completely free from personal bias (a problem which is unfortunately encountered only too frequently in variable star observation).

Frequency of Observation

The variable star observer does not normally concentrate upon one particular class of variable; indeed, for certain reasons which will soon become obvious, the dwarf novae are not the kind of star well suited to the raw beginner to this work, the long period variables being far easier because of their general brightness at maximum and their relatively slow variation. In the case of the long period variables, it is unwise to observe them more frequently than once or twice a week even if conditions are favourable since not only are the daily light variations so minute and difficult to estimate, but personal bias can lead to serious inaccuracies.

By comparison, the dwarf novae, because of their unpredictable, sometimes completely erratic, and often large and rapid, light changes must be observed on every possible occasion. At those times when the variable is caught on the rise to maximum, it is advantageous to make more than one estimate during a single night in order to define the light curve accurately. The small subgroup of the Z Camelopardalis variables present one additional feature, namely the 'standstills' and it is important that both the onset and the termination of these 'standstills' should be observed in order to determine their duration. Mention has already been made of the rare occasions on which a 'standstill' had terminated either in a rise to maximum or even in a rise to a second 'standstill' and since we have no idea, as yet, of the actual physical mechanism of these peculiar methods of comparative inactivity on the part of the variable, it is vital that as much observational information as possible should be obtained concerning them.

Locating the Variable

This is a problem which is encountered in the observation of almost every type of variable star and is particularly acute in the case of the dwarf novae. Not only are they almost invariably very faint objects, even when at maximum but they are also normally white stars, especially when bright and indistinguishable from the large numbers of comparison stars found close by. The long period variables, on the other hand, are all red stars and often stand out from their background by virtue of their colour.

As for all variables, charts are prepared, normally covering 9°, 3°, 1° and 20′ fields, the first showing the position of the variable with respect to naked eye stars and the remainder gradually reducing the area of the field while at the same time introducing fainter and fainter comparison stars with which the brightness of the variable may be compared (these are all stars whose magnitudes are known to be constant and whose brightness has been accurately determined; as far as possible, they are also chosen from among those which are of roughly the same spectral type as the variable itself). Certain of these standard stars are common to two charts, thereby leading the observer step by step, from one chart to the next, going progressively from the naked eye stars to the faintest visible in the field. With most of the dwarf novae (and other variables as well), it is quite possible that when around minimum, they will not be visible in the instrument used by the observer and the problem of locating the variable when it is at the very limit of vision resolves itself into one of recognising star patterns in the field. Fortunately, the human eye is peculiarly well adapted to this and generally there is very little difficulty encountered and what there is, usually disappears as the observer becomes more adept in this kind of work. Very often, however, some uncertainty may arise when trying to relate the field as shown in the telescope with that of the chart and it is helpful to know the size of the field as seen through the telescope, this depending upon the magnification being used and the focal length of the instrument.

The use of setting circles and a driving clock may sometimes be advantageous, particularly with those variables situated in densely populated star fields, but more often than not this method is both tedious and time-consuming. One grave disadvantage is that it requires the use of lights and even if a faint red light is employed, this inevitably reduces the sensitivity of the dark-adapted eye to observing faint objects. Setting circles are very rarely necessary except with large instruments using a high power.

Making Visual Estimates

Once the variable has been located, it is then necessary to make as accurate a visual estimate of the brightness as possible. The two methods generally employed for this are common to the observation of all classes of variable star and since everything depends upon the estimate being as precise as possible, they will be described in detail here.

(a) The Fractional Method: Mention has already been made of the comparison stars which may be used for finding the variable. Their main use, however, is as a sequence against which the magnitude of the variable may be compared. Two comparison stars are chosen, one slightly brighter than the variable and the other slightly fainter. The brightness of the variable must then be expressed as a fraction of the difference between the two comparisons. Since the magnitudes of the comparison stars are accurately known, that of the variable may then be found by simple arithmetic. Occasions will naturally arise when the variable is considered equal in brightness to a particular comparison star and in this case, it should also be compared with at least one other and if more than two suitable stars are available, this method may be used on more than one pair. It may also be that it is not possible to find a suitable star either brighter or fainter than the variable when at either extreme of its range (in the case of the dwarf novae, the former is very unlikely since none of these stars are brighter than eighth magnitude at maximum) and here the dwarf nova should be compared fractionally with two fainter stars of similar magnitude.

(b) Pogson's Step Method: In this method, estimates are made in steps of 0.1 magnitude from as many comparison stars as possible, some being brighter and some fainter than the variable. Although this requires that the eye must be trained to judge such small steps, it does have the advantage that, unlike the Fraction Method, the variable is being compared with only one star at a time. It is also superior to the previous method in that it will enable the observer to detect any variability in a comparison star or a wrongly assigned magnitude and can also be used (although it is not recommended on the grounds of accuracy) when only one comparison star is available.

Since the U Geminorum and Z Camelopardalis variables are such faint objects, there will inevitably be times when too small an aperture or unfavourable seeing conditions result in the star being invisible. If the variable is just at the limiting magnitude of the telescope, a higher power may sometimes darken the field sufficiently for it to be seen, or averted vision may be used since the outer area of the eye are more sensitive to light than the central portion which, as we shall see later, is more sensitive to colour differences. If the variable is totally invisible, negative estimates are extremely important in the study of the dwarf novae since they may be sufficient to preclude the possibility of a maximum having occurred during a period when positive observations are not available.

Limits of Accuracy

One of the main sources of inaccuracy in the early observations of U Geminorum made during the latter half of the nineteenth century, was the widely different, and sometimes quite arbitrary, magnitudes which the various observers assigned to the comparison stars. Quite clearly, unless there are standard sequences in use, the visual estimates made will be far from accurate even without taking into account any personal error on the part of the observer himself. Fortunately the position improved considerably during the nineteen-twenties and such standard sequences of comparison stars with magnitudes obtained photovisually and tied in with the North Polar Sequence, are now readily available.

The so-called Purkinje Effect also plays a part in the production of discordant estimates among observers of variable stars. This is due to the fovea of the eye, when light-adapted, becoming more sensitive to red light than white. Although this can lead to serious differences among the visual estimates made by observers, it is relatively unimportant in the study of the dwarf novae since these are all white stars and theoretically it should be possible for estimates on any one night to agree within 0.2 magnitude.

Sources of Error

One of the main sources of error is mis-identification of either the variable or a comparison star since this leads to results which are both unreliable and misleading. We must remember that all of the dwarf novae are relatively faint, the majority extremely so, particularly at minimum and in crowded star fields it is very easy to mistake the variable.

Bias is another human failing which must be overcome. All previous estimates of the variable and the magnitudes of the comparison stars should be disregarded if possible. One should never assume that the star will continue to brighten or fade in spite of what the previous result would tend to show. Very pronounced irregularities are to be expected in the light curves of these stars and with the Z Camelopardalis variables in particular, the onset and termination of a 'standstill' can only be determined accurately if bias is avoided completely.

The use of averted vision can sometimes be a source of error although when the star is faint it must necessarily be used. Averted vision, in itself, is not a source of error but what must be avoided at all costs is comparing the variable with a comparison star situated at the very edge of the field, a star which will almost certainly be out of focus.

Another problem which has been quite fully investigated in relation to variable star observation is position angle error. Several astronomers have shown that of two equally bright stars in the field, the lower of the two will appear the brighter, as will also that which is to the observer's nose side of the field. The ideal situation is for the observer to bring each star as quickly as possible into the centre of the field when making a comparison.

Finally, we must consider the effect of atmospheric absorption. This is considered to be zero at the zenith and increasing with decreasing altitude, becoming a very serious problem when the variable is close to the horizon. This may be minimised by the use of comparison stars which lie close to the variable, the extinction coefficient then being virtually the same for both stars. If this is not possible, then recourse must be made to the tables of published data and suitable corrections applied. Where the dwarf novae are concerned, there is almost always a suitable number of comparison stars within half a degree of the variable and little difficulty is encountered.

Photographic Observation of Dwarf Novae

We have seen that the U Geminorum and Z Camelopardalis variables are among the faintest of all variable stars and small telescopes are of little use in their investigation except, perhaps, for SS Cygni and its southern counterpart VW Hydri. The others all require instruments larger than 6 inches in diameter for any sizable portion of their light curves to be covered and for observations at minimum, a 12-inch telescope is the minimum.

Recently, the International Astronomical Union at its meeting in Prague has stressed the importance of continuous monitoring of these variables when at minimum phase. The fact that the primary eclipse in the U Geminorum system has an amplitude of one magnitude when the star is at minimum would imply that such light fluctuations may be followed visually provided a large enough instrument is used. Now to what extent may smaller instruments be used in conjunction with photography to study these variables at minimum? Since the limiting photographic magnitude may be anything up to two magnitudes fainter than the visual limiting magnitude when fairly long exposures are employed, this would appear to be an ideal method. Up to a point it is. There are, however, certain restrictions on the use of photography which must be borne in mind as far as a continuous patrol of these variables is concerned. Apart from the cost of the materials, the time taken for one estimate is far longer than that required for a visual determination (at least half an hour as compared with two or three minutes) and this will necessarily restrict the number of observations which may be made in any one night. The long exposures required also rule out the use of this method for following any changes due to an eclipse, the duration of which is shorter than that of the time of exposure. Clearly, where photography is concerned, there are two main fields in which it is invaluable. Firstly, of course, it may be used for the discovery of dwarf novae as well as other types of variable. There are still many areas of the sky which have not yet been fully examined for variables, a fact which is immediately obvious from an examination of the New General Catalogue of Variable Stars; the number of variables discovered in the large constellations along the galactic plane far exceeds those known in similar constellations towards the galactic anticentre. Secondly, long exposure photography, even with a moderate-sized instrument, will provide positive estimates of the brightness of these stars when they are too faint to be detected visually, although as we have mentioned, any rapid variations will inevitably be filled in. Now several of the dwarf novae would appear to have fairly large colour indices and if such estimates are to be used in conjunction with the visual light curves, they should preferably be made using the appropriate filter and photovisual plates.

Photoelectric Estimates

These are, of course, the most accurate, particularly in following rapid light variations and detecting the presence of eclipses in these systems. Once again, large apertures are necessary in order to measure the small light changes in an object which is extremely faint at minimum phase. Since photoelectric cells differ widely in their sensitivity to different colours, there is no standard photoelectric scale of magnitudes. Within certain well-defined limits, the photoelectric current is directly proportional to the intensity of the incident light and this gives it an immediate advantage over the photographic plate since the time taken to make one reading is extremely small and the accuracy which can be achieved, often up to ten times better than that from photographic plates, enables us to measure very small changes in brightness. It is, however, necessary to compare the variable with a comparison star having a spectral type as similar as possible to the dwarf nova.

15. Cataclysmic Variables

It is now time to give a very general summing up of the information given in this book. It will have become clear that the cataclysmic variables, and in particular the dwarf novae which have been the main subject of discussion, show many features in both their light and spectroscopic variations which are completely different from those of the other classes of variable stars. During the first few decades of their investigation when only a dozen or so were known, quite arbitrary mathematical relationships were put forward in an attempt to explain the irregular variations in their mean cycles, most of these aimed at showing that the various lengths of the intervals between outbursts followed certain rules in their succession or that the amplitude could be correlated in some way with the length of the mean period. Later, as more of these peculiar variables were discovered and many more observations became available for discussion, an empirical relation was found between amplitude and mean period which was more or less linear. Yet the data were still few and there were some stars which, although beyond all doubt belonging to the dwarf nova class, did not fit the curve at all. In spite of these serious limitations, the cycle-amplitude relation did have one important result, namely that it showed a close family relationship between the dwarf novae on the one hand and the recurrent and ordinary novae on the other.

Since the discovery of U Geminorum in 1856, there had always been, of course, the obvious fact that the abrupt increase in brightness of between two and five magnitudes, sometimes occurring in a single day, pointed directly to a close connection between the three classes of cataclysmic variables. The general characteristics of the light curves, too, were similar if allowance was made for the wide differences in mean periods. It may be said that, at the beginning of the century, the opinion was widely expressed that the basic difference between the dwarf novae, recurrent novae and ordinary novae was one of degree and not of kind, with the amplitudes of the three types fitting in extremely well with the derived mean cycles. Certainly, the picture presented indicated a logical sequence and although it gave no physical explanation of the peculiar light changes apart from the fact that they clearly seemed to be explosive in nature, it was considered as a basis on which to formulate various theories of the constitution of these stars.

It is interesting to speculate on how much additional information would have been added to our knowledge of the nature of these stars had it been possible only to observe their light variations. We have, of course, the classic example of the Algol variables for which the correct explanation of their light changes was given by Goodricks long before it could be proved by more refined techniques. These eclipsing binaries, however, have periods which are usually completely regular, often to within a second or so; a condition which is far from true for the dwarf novae. It is indeed doubtful if the early investigators would have been able to gain much additional information concerning these eruptive variables from the light curves alone and certainly, any attempt to deduce a relation between the maximum brightness of any particular outburst, or its duration, with the length of the minimum either immediately preceding or following it, is doomed to failure.

With the eventual recognition of the small Z Camelopardalis subgroup, a further complication was added to the already bewildering array of puzzling features associated with the dwarf novae. The phenomenon of the 'standstills' presented astronomers with a problem which is even now, far from solved. Why the star should suddenly halt in brightness on the decline to minimum and remain at an intermediate magnitude for a period which may be anything from a few days to more than a year is enigmatic to say the least. Clearly, speculation concerning this discontinuity in the fading of these variables after certain maxima will be unprofitable until we know the cause of the outbursts themselves in the Z Camelopardalis stars.

The application of the spectroscope to the problem was a

natural consequence of the pioneering work of Huggins and came fairly early in the study of these variables and as long ago as 1882, the presence of bright emission lines in the spectrum of U Geminorum was demonstrated, apparently adding further confirmation to the belief that the star is similar to the novae. As techniques became more refined and larger instruments came into being, better spectra were obtained but still having a relatively low dispersion. Instead of clarifying the position, these spectra only served to introduce further puzzling features which were incapable of resolution at the time.

One important fact which did emerge, however, and which had a profound bearing on future developments, was the behaviour of the emission lines. Several bright novae had been examined spectroscopically shortly after maximum brightness and in almost every case the emission lines were found to be displaced towards the violet end of the spectrum indicating that the source is moving towards us, the actual velocities found being extremely high, of the order of 1,000 kilometres per second. Clearly, a shell of gas consisting mainly of hydrogen is ejected from the nova following the initial outburst and this interpretation may be further confirmed in many cases by both visual and photographic observation once the nova has faded sufficiently for detail to be made out.

The dwarf novae on the other hand show no displacement of the emission lines and it gradually became evident that, whatever the cause of the nova-like outbursts may be, it certainly is not an explosive ejection of gas of the kind found in the other two classes of cataclysmic variable. Here the matter rested for many years until high-dispersion spectra taken with the 100- and 200-inch reflectors provided the first clue to the nature of these stars. The composite spectrum of RU Pegasi at minimum immediately revealed it to be a binary system and in the years which followed, more dwarf novae were shown to be binaries. Once this duplicity had been established, not only for the U Geminorum and Z Camelopardalis variables, but also for the novae and recurrent novae, other vital details soon came to light.

All of the cataclysmic variables examined are in an advanced evolutionary phase since every one contains a component which is a hot subdwarf star closely akin to the white dwarfs and presumably one in which helium burning has reached an advanced stage. The late-type components do not form so homogeneous a group although those of the dwarf novae form a much more compact group than those among the recurrent and old novae, all being subdwarf stars with a spectral type later than dG3. Other similarities were also found in these systems. In every one, the late-type star fills its inner surface of the equipotential surface as defined by the restricted three-body problem and there is ejection of mass through the inner Lagrangian point into the lobe of the blue companion. This material takes up an orbit around the white dwarf star, forming a quasi-permanent ring which is not uniform either in density or brightness, the brightest portion apparently lying between the inner Lagrangian point and the following hemisphere of the blue component.

Although the number of dwarf novae bright enough for satisfactory high-dispersion spectrograms to be obtained is small, it is sufficiently large for us to expect that one or two will show evidence of eclipse. Such variables have indeed been found, these being U Geminorum which is a single-lined spectroscopic binary and Z Camelopardalis which is even more valuable since it is a double-lined spectroscopic binary and is the one dwarf novae for which we can make direct estimates of all of the parameters of the system. The calculated masses of 0.7 and 0.8 solar masses for the blue and red components respectively are in excellent agreement with those of the other dwarf novae which have been estimated by more indirect methods.

The investigations of the past twenty years or so have thus provided us with a fairly clear picture of a dwarf nova system – a short-period binary consisting of a hot, blue subdwarf star and a much cooler and slightly more massive companion which fills its lobe of the equipotential surface and is ejecting mass through the inner Lagrangian point, thereby forming a semi-stable ring around the blue star. That the ring of gas lies in the plane of the orbit is shown by the presence of doubled emission lines in the two systems which are known to be eclipsing variables, the doubling being due to the familiar Doppler shift originating in the approaching and receding edges of the rotating ring.

Such a model of the dwarf novae is now generally accepted but, of course, it does not explain the typical nova-like outbursts. Until a few years ago, these were believed to originate either in the hot, blue star or some region very close to it, one further tentative suggestion being that the lobe of the equipotential surface surrounding the blue star is filled with a very tenuous medium which somehow changes in opacity, possibly due to electron scattering, giving rise to the abrupt increases in brightness observed in these systems. The fact that the late type component might be the seat of these outbursts was not even seriously considered until the very accurate photoelectric investigations carried out by Krzeminski on U Geminorum. This has now proved almost conclusively that, in this system at least, it is the red star which is responsible for the eruptions. On this theory, the late-type star possesses a convective envelope and irregular oscillations can occur on the pulsational time scale. These result in an asymmetrical expansion of the outer atmospheric levels, an associated increase in the mass loss through the inner Lagrangian point with a subsequent increase in the surface temperature of the star due to a baring of the lower photospheric levels.

It is clearly impossible to overestimate the importance of this discovery to the study of the dwarf novae as a whole. Although it would be unwise at present to assume that this sequence of events takes place in all members of this class, there would certainly appear to be no observational evidence against such a hypothesis. It also has the very distinct advantage of removing the problem we encountered earlier of the absence of any Doppler shift of the emission lines following an outburst as found in the novae (in these latter stars there seems no doubt at all that the nova phenomenon is due to an explosive outburst of the blue component).

Our knowledge of these peculiar stars has clearly advanced a great deal since the speculations of the early investigators, much of this vital information having been gained by the use of the 100- and 200-inch reflectors using very refined spectroscopic and photoelectric techniques. Such facilities will doubtless be used in the future to tackle the outstanding problems still remaining. We have now clearly reached a point where we have a reasonably firm base on which to build. The discovery that the red component is the seat of the eruptions in the U Geminorum system has obviously gone a long way towards a detailed explanation of the physical events which lead up to this phenomenon. But how far does it enable us to explain the curious 'standstills' of the small subgroup of the Z Camelopardalis variables? Here we still have some way to go before a wholly satisfactory answer is found. Most of the clues are already there – the fact that for each individual star the 'standstills' appear to begin at virtually the same position on the decline to minimum; that their duration is not related either to the maximum brightness attained by, nor the the total radiation emitted by, the preceding maximum; and that the apparent constancy of brightness throughout even the longest 'standstill' would suggest that somehow, the mechanism responsible for the normal outbursts, is suppressed during the duration of a 'standstill'. It seems very likely that if we are to learn the physical nature of the 'standstills', the information will come from accurate photoelectric investigations of Z Camelopardalis itself since this is the only member of this small subgroup which is known to be an eclipsing binary. Whether the fact that RX Andromedae is the only dwarf nova known with an eccentric orbit is significant in this respect is problematical at present.

The second outstanding problem relating to these variables is that of evolutionary career. The earlier view that they are closely related to the novae and recurrent novae clearly has several points in its favour. All three classes of star are binary systems and all contain a hot subdwarf as one of the components. The same empirical cycle-amplitude relation which has been found for most of the dwarf novae also satisfies the recurrent novae and possibly, by extrapolation, the novae, too, although their mean periods are measured in hundreds or thousands of years. Certainly the presence of strong emission lines of ionised helium in the spectra of the old novae (and recurrent novae) is indicative of a much higher surface temperature for the blue components and the nature of the outburst is completely different. Nevertheless, these arguments do not rule out a close relationship by any means. A much more serious argument against this hypothesis lies in the spatial distributions of the dwarf novae as compared with the other two classes of eruptive variables. Whereas the novae are virtually all members of the galactic disc and not of the galactic halo, the dwarf novae appear to belong to a population intermediate between the two; almost as many are known in the region of the galactic poles as there are in the Milky Way.

Several lines of argument are now being developed to show that the W Ursae Majoris eclipsing variables are the most probable progenitors of the U Geminorum and Z Camelopardalis stars. The spatial distributions of the two groups are very alike, their mean periods, the sum of the masses of the two components, the dispersions in their radial velocities and mean distance above or below the galactic plane are all virtually identical. What is perhaps more important, however, is that both the primary of the W Ursae Majoris variables and the red component of the dwarf novae systems are pronouncedly underluminous for their mass.

The W Ursae Majoris variables are all F or G-type stars lying on the main sequence and have a mass ratio of between 2:1 and 3:1 and theory predicts that this being so, the more massive component will commence to evolve away from the main sequence before any evolution of the secondary takes place. Consequently, the primary will expand and once it encounters the critical surface of its lobe it will start losing mass, either into space where it will be totally lost to the system, or by transference to the secondary through the inner Lagrangian point while at the same time it evolves into a blue subdwarf similar to the white dwarfs found in the U Geminorum variables. Now the effect of mass loss by either of these two methods has been very extensively studied and we know that, provided certain conditions are observed, loss from the system will result in an increase in the orbital period and loss to the secondary component will lower the orbital period.

If the dwarf novae are, therefore, descended from the W Ursae Majoris systems it follows that the predominant mode of transfer is to the second component, observational confirmation of this being given by the fact that the sum of the masses is virtually the same in both types of system. The orbital period of a typical W Ursae Majoris star will therefore decrease until by the time the masses of the two components are almost the same (as found for the dwarf novae), the orbital period will have diminished to the average figure found for the U Geminorum variables. At this point in its evolutionary career, the secondary will have begun to evolve, a process possibly accelerated somewhat by the accretion of mass from the primary and it too, will then expand, the role of primary and secondary now being reversed and there is a mass flow in the other direction, back to the blue star. It is at this point in the evolution of these close binaries that we find the dwarf novae now.

Naturally, this theory has had to face up to certain criticisms. The discovery of W Ursae Majoris variables in some of the intermediate-age clusters where they are also main sequence stars would indicate that the age of these variables is at least that of the other main sequence stars in the clusters and suggests that they are far too old to be progenitors of the dwarf novae. More recently, it has been suggested that the later stage in the evolution of a W Ursae Majoris variable will not be a binary of the dwarf nova type, rather the primary will engulf the secondary component completely, forming a single object. How far future observations will go towards deciding for or against the evolution pattern of a generic link between the W Ursae Majoris stars and the dwarf novae is difficult to assess at the moment. A lot is going to depend upon the pre-W Ursae Majoris state for once we understand this more fully, it is certain to influence our ideas on the later stages of the evolutionary development of these contact binaries.

Index